'I have studied many philosophers
and many cats:
The wisdom of the cats is infinitely
superior'

– Hippolyte Taine

'Cats seem to go on the principle
that it never does any harm to ask for
what you want'

– Joseph W. Krutch

'It is my conviction that cats never forget
anything . . . a cat's brain is much less
cluttered with extraneous matters
than a human's'

– Paul Corey, *Do Cats Think?*

Melissa Miller's Definitive I.Q. Test for Cats
and
I.Q. Test for Cat Owners

A SIGNET BOOK

SIGNET

Published by the Penguin Group
Penguin Books Ltd, 27 Wrights Lane, London w8 5tz, England
Penguin Books USA Inc., 375 Hudson Street, New York, New York 10014, USA
Penguin Books Australia Ltd, Ringwood, Victoria, Australia
Penguin Books Canada Ltd, 10 Alcorn Avenue, Toronto, Ontario, Canada m4v 3b2
Penguin Books (NZ) Ltd, 182–190 Wairau Road, Auckland 10, New Zealand

Penguin Books Ltd, Registered Offices: Harmondsworth, Middlesex, England

First published 1992
7 9 10 8 6

Typeset by DatIX International Limited, Bungay, Suffolk
Set in 10½/13 pt Monophoto Sabon
Printed in England by Clays Ltd, St Ives plc

For my sister, Laura,
and her cat, Champagne,
who were the inspiration for
much of this book

CONTENTS

viii

ACKNOWLEDGEMENTS

I would like to thank Celia Haddon for her encouragement throughout this project and my former colleagues at the WM Company in London for their patience and flexibility while the book was written. Thanks are also due to over sixty-five cat owners who tested their cats' I.Q. as a contribution to my research. As ever, I am grateful to my family and friends for their wholehearted enthusiasm and unerring support upon which I can always rely.

For their permission to quote from copyright material, I would like to thank the following: HarperCollins Publishers, Glasgow, for a quotation from the *Collins Concise English Dictionary*, edited by Patrick Hanks (first edition 1982); The National Geographic Society for an extract from the April 1964 *National Geographic* magazine article, 'The Cats in Our Lives', by Adolph Suesdorf (ref: 125:508); Adriaane Pielou for extracts from her 11 November 1990 *Mail on Sunday* article '100 Ways to Cure a Cat'; Sidgwick & Jackson, London, for passages from *Do Cats Need Shrinks?* by Peter Neville (Copyright © Peter Neville, 1990); Random Century Group, London, and Crown Publishing Group, New York, for passages from *Catlore* by Desmond Morris (Copyright © Desmond Morris, 1987); André Deutsch Ltd for an extract from 'Cult of the Cat' by Patricia Dale Green in *Nine Lives*, compiled by Kenneth Lillington (1977); Octopus Publishing Group for extracts from *Encyclopaedia of the Cat* by Angela Sayer (1979); Michael Joseph Ltd for *Charles: The Story of a Friendship* by Michael Joseph (1943); Nevill Coghill and Penguin Books for lines from *The Canterbury Tales* by Geoffrey Chaucer.

I regret that, despite every effort, I have been unable to find the copyright holders of *The Siamese Cat* by Helen and Sidney Denham; of selected quotes from the *365 Page-a-Day Cat Calendar*, Workman Publishing Co., New York; of 'O Little Cat with Yellow Eyes' by Helen Vaughan Williams. Should the copyright holders contact the publisher, acknowledgement will be made in future printings.

Lastly, I would like to thank the following cats and their owners who have written to me with their results and stories since this book was first published. The list includes all letters and results received up to the end of April 1994.

From Great Britain: Albert, Alli, Arthur, Bagpuss, Baldrick, Bastet, Brigid the Brown Blot, Coco, Cressida, Cyder, Demi, Dennis, Dusky, Edie-Puss, Fat Cat, Flora, Jessie, Joseph, Katy, Kanya, Kitty, Kitty (Liverpool), Kiwi, Leo, Maisie, Max, Mick, Minty, Orlando, Paddy, Pepsi, Perdita Mew, Philomena, Pushkin, Rover (Plymouth), Rover (Cambridge), Rhubarb, Rum, Scallywag, Sherry, Simba, Spratt, Sultan, Tabitha, Tara, Thomas, Twiglet, Tyrone, Victoria, Wee Cat. *From the United States:* Abby, Alaric T. Visigoth, Amber Scampi, Angela, Antoinette, Bear, Bella, Bomb Abominable, Boots, Bunny, Butch, Buttons, Cameo, Cecilly, Charlie, Charlie Dickens, Chester, Chester (Los Angeles), Cinnabar, Dory, Frost Wizard, Kitty-Boo, Lucy, Mischief, Miss Dickens, Oliver, Pasha, Peanut, Perly, Pickle, Pooh Bear, Rim Shot, Rhiannon, Roberta, Rockit, Rocky, Rolex, Sappho, Scooter, Selena, Senator Noodles, Sibby, Tabitha, Tasmanian Devil, Tiger, Tika, Zack. *From Canada:* Em, Isobel, Murray, Rover. *From France:* Guismo, Samuel. *From Guam:* Barney, Bella. *From Ireland:* Lilly. *From New Zealand:* Rajah. *From Switzerland:* Bilbo Baggins, Cobber, Rhian.

I
INTRODUCTION

———

I have studied many philosophers and many cats:
The wisdom of the cats is infinitely superior.

No intelligent person can be conceited in the presence of a cat
No living creature can so ignore.

· Hippolyte Taine (1828–93) ·

THROUGHOUT the five thousand years that the cat has lived in a domesticated setting, man has speculated upon the level of intelligence it seems to possess. Given the cat's ability to reason and to think for itself, as well as its air of utter superiority and rather mysterious nature, it is perhaps easy to understand how it has captured the imagination of so many generations.

Anyone who has ever lived with a cat would probably agree that they are, in general, quite clever animals. They are adept at analysing situations and are usually quick to learn – how to switch off a light, for example, or how to open a door, most often of a food cupboard or refrigerator. Cats are also keen observers of everything that goes on around them, able to devise well-planned strategies to catch their prey when hunting and to interpret our moods and feelings at almost any moment.

Indeed, there are many examples of intelligent behaviour in the average domestic cat to convince most observers of its excellent mental faculties. Speculation on intelligence is encouraged by the cat's mysteriousness; it is often difficult to know exactly what a cat is thinking. They can be very expressive but they do not always share their feelings. Many enjoy quietly perusing their environment like old sages, leaving us to wonder what they might be contemplating. Their independent nature contributes to their mysterious image, and we may not always understand the motivations for their behaviour, or sometimes even know where our cats are or what they're doing.

Cats also appear to possess the ability to predict certain events and many people believe that they have extra-sensory perception. In the case of cats, it might be more accurate to call it *excellent* sensory perception, as it is their phenomenal ability to see, hear and detect changes in the atmosphere which enables them to react well before we recognize that anything is happening.

Because of the cat's clever behaviour, it is tempting to credit it with more intelligence than it may have, and the cat has been honoured with one of the largest collections of myths and legends that exists about any animal. The Ancient Egyptians, for example, believed that cats possess supernatural powers which were bestowed upon them by the gods. As the animal representative of Bastet – the goddess of fertility, happiness and the moon – cats were thought to be capable of influencing the fertility of human beings, as well as that of the land. This link is appropriate as cats are fertile creatures themselves. They are also well-placed as symbols of Bastet's affiliation with the moon since cats' phosphorescent eyes were assumed to work on the same principle as the moon at night, reflecting light 'stored' from the sun.

The association between the cat and the moon is a recurring theme in legends. In Roman mythology, cats represent Diana, the huntress-goddess and the goddess of the moon. Both Diana and the cat are further identified with Hecate, originally a moon goddess in Roman mythology, who later became the goddess of the underworld. Hecate possessed evil supernatural powers which the cat was subsequently credited with, such as the ability to transform itself into a witch and to enchant human beings.

One of the greatest promoters of such myths about the cat was the Christian Church in Europe during the Middle Ages. When its power was threatened by a disgruntled populace, the Church blamed an epidemic of demonic witches, along with their cohort, the cat, for the ills of the time. Across Europe and in colonial North America, all levels of society were eventually convinced that cats posed a threat to mankind. Sir Walter Scott, for example, believed that their minds were filled with menacing thoughts, claiming that 'it comes no doubt from their being too familiar with warlocks and witches'.

For over three hundred years, cats were persecuted as servants of the Devil and as accomplices to the many 'witches' who were put on trial and, almost inevitably, declared guilty. By the late nineteenth century many of the old myths about the cat, including its connection with witchcraft, were being dispelled by the objective and rational developments of science. Research has determined why, for instance, some of the myths may have developed in the first place. The proven ability of the cat to hear ultrasonic sounds and to detect changes in the atmosphere – which enables it to predict changes in the weather as if it had ESP – is just one case.

In some cultures this talent in cats has been recognized and used to man's advantage. The Chinese government, for example, began monitoring the behaviour of cats and other animals at a number of their seismographic centres in 1970, looking for signs of unusual behaviour which could indicate that an earthquake was on its way. In 1975 cats in the central city of Haicheng were seen to be acting strangely; the entire city was subsequently evacuated and the area was devastated by an earthquake the next day.

But while science has dispelled many of the myths about the cat, it has at the same time confirmed that cats are in fact quite intelligent. Many animal psychologists find the cat fascinating to study because its brain is considered to be the most similar to the human brain in its structure and functionality. And in laboratory experiments, specifically designed to test cat intelligence, cats have performed consistently well and exhibited an impressive ability to reason and to learn.

Many cat owners suspect that their cats are intelligent without any scientific evidence to confirm it. The domestic cat has demonstrated its high level of intelligence in the way it has created a niche for itself in modern households. Cats manage to receive provision of food, shelter and other comforts without the guarantee of giving anything in return. Unlike dogs, who guard our homes and volunteer their unerring loyalty in exchange for the benefits of domestic life, cats generally know how to get what they want without sacrificing much themselves. Indeed, cats somehow make us feel that it is *their* choice to live with us and should they wish to return to the wild, they would be quite able to care for themselves!

With this book, cat owners now have the opportunity to quantify the intelligence of their favourite feline and confirm or disprove any suspicions about its mental abilities. Owners answer each question on behalf of the cat, drawing largely upon past observations of its behaviour. But the cat needs to be present for some of the questions which test its spontaneous reaction to certain stimuli and situations.

This I.Q. test is meant to be accurate but, above all, amusing and entertaining to take. Questions are grouped into four areas: visual skills, audio abilities, social behaviour and domestic behaviour, and each question has been designed to test a cat's capacity for reasoning and intellectual thought.

The test measures several of the same areas of mental ability as today's I.Q. tests for humans, including verbal comprehension (your cat's response when its name is called), vocabulary (breadth and understanding of cat sounds), memory (recognition of the cat carrier, and what happened the last time you brought it out), spatial ability (judging the height of kitchen cupboards), and reasoning ability (how to cajole extra food).

A cat's total score can be translated into a cat I.Q. using a conversion graph within the results section. The average I.Q. for cats is equal to 100, determined from the average score of a representative sample of domestic cats, each of which was tested by its owner, using the test in this book, and which are named in the Sample Pool List of Participants at the back of the book. Your cat's I.Q. can thus be ranked against this sample of the domestic cat population to determine whether it is Blissfully Ignorant, Occasionally Quite Clever, Average, Brighter Than Average, Very Intelligent, Extremely Intelligent or Possibly Brighter Than You Are.

But what about your own I.Q. as a cat owner? Relationships between cats and their owners can be complex, as the growing popularity of cat therapists and cat psychologists testifies. Some say that it is not the cat which has been trained by its owner, but rather the other way round. How often does your cat get its own way, apparently never doubting that it would? Many owners delight in spoiling their cats, while some still try to maintain an upper hand around the house.

With this book you can also measure *your* level of intelligence in the role of cat owner. Like the I.Q. test for cats, a greater emphasis has been placed on amusement rather than on accuracy, but the test should give you an insight into your relationship with your cat.

The test has been designed to measure your efforts in the relationship, the responsibilities you assume, and the attitudes and feelings you have about your cat. After a brief series of background questions which examine your past experience with cats, the test is divided into three further sections which assess your dedication, sensitivity and the degree to which you have been able to train your cat as a pet – as well as the reverse!

Once you have answered all the questions and added up your total score, you can refer to the Results Analysis section to convert it to a cat owner I.Q. You can then consult the Cat Owner I.Q. Classification Table to determine what type of owner you are. The highest scoring category is that of the Fanatical owner, one who dotes on his or her cat and gets as much pleasure from spoiling it as the cat does. The second highest scoring category is the Congenial cat owner, one who enjoys treating his or her cat very well, but is slightly wary of it becoming too spoiled. In the third category, Flexible, owners are more rational in the treatment of their cat, but give in to their cat's wishes more often than they may realize. The fourth and lowest scoring category is that of the Practical owner, one who appreciates the presence and companionship of a cat, but does not want to be bothered with one that requires excessive maintenance or attention; indeed, the more self-sufficient the cat is, the better suited it will be. Detailed descriptions of the ideal cat and recommended cat breeds for each type of owner are included after the test.

At the end of the book is a one-page questionnaire which asks for the I.Q.s of both yourself and your cat, as well as other relevant information such as the number of years you have kept a cat and your cat's age and breed. The author is collecting this data for a follow-up on the I.Q.s of cats and their owners from around the world; if you would like to contribute to the research, she would be most grateful.

As you take the tests, have fun and do remember that the success of the relationship between a cat and its owner depends on a great number of factors – your respective I.Q.s play a very minor role in the happiness you can give one another.

II
THE CONCEPT
OF CAT INTELLIGENCE

INTELLIGENCE is an abstract concept very difficult to define. In dictionaries it is most often said to be 'a capacity for understanding' or 'an ability to comprehend and perceive meaning'. If the intelligence of an animal is defined as its ability to understand a concept or a subject, the *degree* to which the animal understands should correspond directly to the degree of intelligence the animal possesses.

Levels of intelligence are easier to detect in humans than in other animals. Humans can explain their understanding of a concept through speech or the written word, indicating their level of intelligence. When considering the intelligence of other animals, it is necessary to look for evidence of intellectual thought primarily by observing their behaviour. Isolated instances are fairly easy to identify as, for example, when an animal confronts a situation, analyses it and then solves it. But to quantify an animal's entire *capacity* for intellectual thought is a difficult exercise.

THE INTELLIGENT CAT IN SCIENCE

Intelligence tests for humans were first devised in 1905 by the French psychologist Alfred Binet. His tests were originally designed to measure the aptitude of children, so that those who were too slow to benefit from formal education could be given special remedial classes by the French government.

Binet's test contained thirty questions which by modern standards would be considered biased since they placed a heavier emphasis on the amount of general knowledge a person already had than on the ability to judge and reason.

To improve I.Q. tests for humans, it was suggested that the general knowledge element be removed and that intellectual capacity be tested by a wide variety of problems and mental tasks. After all, the purpose of the test was to define one's ability to comprehend, analyse and reason, not to measure the amount of stored knowledge.

Thus I.Q. tests were subsequently refined and expanded so that a person's mental acumen could be measured more accurately. To avoid evaluating one mental attribute above another, I.Q. tests eventually addressed seven different areas of mental ability. These are still used in today's I.Q. tests for humans and include reasoning ability (completing a series, for example), spatial ability (reproducing designs by memory), perception (speed in comprehending visual details from pictures), memory, vocabulary, verbal understanding, and numerical skills.

To determine the I.Q. of the average domestic cat is a bit trickier than determining the I.Q. of a human for, firstly, it cannot express itself in words. Secondly, an I.Q. test for cats needs to cover as wide a range of mental abilities as the I.Q. test for humans, but it also has to be tailored to the type of activity in which cats can participate. Word fluency and numerical skills, for example, would not be appropriate — at least not in a general test designed for the *average* domestic cat!

In this book questions are divided into four general areas of visual skills, audio abilities, social behaviour and domestic behaviour. Each question is designed to test the reasoning ability of a cat as observed by its owner in everyday situations. The mental abilities which are measured in I.Q. tests for humans and included in the I.Q. test for cats are memory (your cat's response when shown a noisy electric appliance), vocabulary (repertoire of cat sounds), spatial ability (ability to descend or ascend to different levels), verbal comprehension (your cat's response when given a command) and reasoning ability (your cat's reaction if you hide its food behind your back).

Once all the questions have been answered and the cat's total score established, this can be converted to a Cat I.Q. using a graph within the results section. As in I.Q. tests for humans (where an I.Q. of 100 represents the average score from a sample of the population), the average I.Q. for cats is equal to

100 which reflects the average score from a sample pool of domestic cats.

Of course some say it is inaccurate to compare an individual's score with a large sample pool because of the wide variety of ages, educational backgrounds and other characteristics represented which affect the score of each pool member. Given the diversity of lifestyles and backgrounds within the human population, this may be true for I.Q. tests for humans, but it is less of an issue within the domestic cat population.

The accuracy of human I.Q. tests does come into question as a person can score differently on the same I.Q. test when tested at different times. He or she may deserve a higher score, but simply does not test well or was nervous or sick on the day of the test. In the Definitive Cat I.Q. Test it is the owner who answers each question on behalf of his or her cat, choosing when to do the test and spending as much time on it as he or she pleases.

Furthermore, most answers are based largely on the owner's past observations and experience of his or her cat which will, by definition, have been completely natural and uninhibited displays of intelligence (or the lack thereof!). It should be remembered, however, that an I.Q. is no different from any other form of physical measurement in that it does not represent what is measured in and of itself, but merely gives an indication of the dimensions of what has been measured.

Defining a cat's I.Q. is just one way of determining how intelligent a cat may or may not be. Other methods have also been used to test and quantify cat intelligence, mostly in controlled laboratory environments. In almost every experiment or series of tests, it has been confirmed that the average cat is indeed quite intelligent.

When it comes to testing cats in a laboratory setting, though, cats are particularly difficult to analyse because of their independent natures and inconsistent responses to standard types of experimental stimuli, such as winning a reward on offer.

Cats are able to analyse the strategies available to them and will choose the one which requires the least amount of work.

An experiment which successfully showed cats' intelligence was a test involving puzzle boxes, where a cat was put inside a closed box and had to discover its way out. Not surprisingly, this was something each cat did indeed want to do. A variety of boxes were used which were increasingly difficult to escape from. In almost every case, however, the cat was able to determine its way out of the box, including the most challenging with complex hooks and catches on the doors. When put back inside the box, the cat was able to remember how to get out and would do so quickly even if it had taken it some time to get out in the first instance.

In a different series of tests conducted at Wesleyan University in Connecticut, cats were offered food temptingly suspended on a piece of string and out of reach. A box on wheels – with which the cats had already become familiar and had learned to pull around the room – was placed close by. After trying the conventional methods of jumping up from the floor or leaning from the top of the nearby box, the cats were able to reason that by pushing the box directly under the string, they could then easily reach the tantalizing food.

THE INTELLIGENT CAT AT HOME

Evidence of feline intelligence can easily be found beyond the laboratory, as anyone who has lived with a cat will agree. Curiosity, for example, the well-known indicator of intelligence, is exhibited by many domestic cats. They enjoy frequent investigations of their surroundings, even if these do not change at all.

Cats have been known to learn how to open doors and windows, and sometimes even the refrigerator. They can recognize familiar sounds, especially favourite ones such as a can of their food being opened or *your* food being unwrapped. Some cats can learn to operate machines for their own amusement, knowing which button to push to turn on the radio, for example, or how to play and eject a tape from a video-recorder.

In general, cats are also able to recognize and adapt to the behavioural limits set by their owners. They can usually be trained to use a litter box, particularly if taught at an early age; the majority will also learn not to jump on to kitchen surfaces and the like, unless the owner does not mind. Clever cats who are not supposed to do so will only jump on to work tops when their owners are not around!

Without the help of an alarm clock, many cats seem to know instinctively when it is time for their owners to wake up, the time for their regularly scheduled meal, or when it is time for their owners to return home when they can be seen waiting by the door. Cats are very astute observers of all that goes on around them and are sometimes able to remember daily routines better than their owners.

In the case of one bright cat, for example, its owner had a morning routine of waking up, making a cup of tea and then feeding the cat. When the person overslept one day, missing

not only breakfast but also lunch, the cat pulled an old tea-bag from the kitchen bin and promptly deposited it on its owner's pillow. It had made the connection that a cup of tea for its owner meant a plate of its food would soon follow for it.

Other cats can sense when their owners are to go on holiday, apparently understanding what's about to happen when they see a suitcase being packed or, as if by intuition, acting upset even before there are any signs that their owner will soon be leaving.

Masterful hunters, cats are expert at concealing themselves as and when necessary, and can skilfully assess their surroundings to determine the best line of attack. 'I have watched our young ... Manx carefully avoid a dry leaf while stalking, apparently because she knows its crackle or rustle might warn what she was creeping up on,' writes Paul Corey, a dedicated cat-watcher, in his book *Do Cats Think?* 'This is such a fine point in the art of hunting that it seems to me to emphasize the cat's fantastic ability to learn.'

Some cats even lay traps for their prey. Cats have been known to take bits of bread into the garden to attract birds and then hide themselves near by until the moment is right to strike. Another documented strategy calls for bread to be strewn over a certain area of the garden and for the cat to sit very still, right in the middle of the bread; it is not clear, however, whether the cat is extremely confident or simply too stupid to conceal itself!

When it comes to judging spatial distances, cats are again adept. They rarely misjudge when they jump on to things or, indeed, when they dive underneath a surface. An old English proverb states that if you 'throw a cat over a house ... it will land on its feet'. Certainly when cats have the misfortune to fall from a height, almost all will land on their feet, unscathed – unless the height is just too great. This indicates an impressive ability to gauge distance at lightning speed, not to mention a flexible body!

THE INTELLIGENT CAT'S SUPERIOR PHYSIQUE

Cats have developed many physical characteristics which enhance their abilities to perceive, comprehend and sense what is happening around them. In many cases these attributes are superior to those found in man. Inasmuch as intelligence is defined by the ability to understand one's environment, such physical advantages as superior hearing and eyesight, for example, may contribute to the cat's level of intelligence and to our belief that most cats are very clever.

If man could be crossed with the cat it would improve man, but it would deteriorate the cat.

· Mark Twain (1835–1910) ·

Unlike humans, for instance, cats can see in almost total darkness. They are able to see in light five times dimmer than the faintest light we can see in because of a superior ability to contract and expand the pupils of their eyes. Hence the cat's ability to move around confidently at night.

As experienced hunters, cats have an acute visual response to even the slightest movement. They have an overlap in their fields of vision which gives them stereoscopic sight. This enhances their judgement of distance whenever they need to jump or pounce on to something.

Cats can also stare for hours without blinking; this can make them look more intelligent than man, who is less capable of scrutinizing objects.

Cats have far more complex and more highly developed ears than ours. Their ears contain thirty different muscles whereas ours have a mere six. They can also hear sounds more than two octaves higher than the highest note we can hear.

Sound travels in waves and humans in their prime can hear sounds at up to 20,000 cycles per second. Dogs can hear higher sounds, detecting sound waves of 40,000 cycles per second, but cats can hear sounds which are speeding by at more than 65,000 cycles per second and, in some cases, up to 100,000 cycles per second. Incidentally, the squeaks a mouse makes are also emitted up to this level, inaudible to humans but within the range of a cat's sharp hearing faculties.

While they appear to be asleep, cats monitor all that is happening around them and can react to something you were sure they would not hear. This can be seen in the cat which looks sound asleep but which suddenly springs to action the moment the refrigerator door opens. According to Joan Hendricks of the University of Pennsylvania, cats are able to 'sort through tremendous amounts of irrelevant noise and pick up significant sounds when they're sleeping'. This is why, she explains, cats 'can appear sound asleep with washer or vacuum running, but perk up at the sound of a can being opened in the kitchen'.

When they are young, cats hear especially well. If two sounds come from the same direction but from different distances, for example, they can distinguish between them. If two sounds are the same distance away, but just eighteen inches apart from each other, they can differentiate between them from a distance of sixty feet. They can also detect the difference in two notes which are less than a semitone apart.

Likewise, cats have a superior sense of smell which helps them to sniff out and understand their surroundings. The average human's nose has approximately nine million specialized nerve receptors with which it recognizes smells. The average cat's nose, however, has about one hundred and forty million.

It is my conviction that cats never forget anything . . .
a cat's brain is much less cluttered
with extraneous matters than a human's.

· Paul Corey, *Do Cats Think?* ·

The cat is often used in experiments which analyse the mechanics of the human brain because its brain is so similar in functionality and structure. In the brain of the cat the most recently evolved areas form part of the cerebral cortex, the largest and most complex part of the brain which surrounds all the older areas. This controls many different functions including the more demanding mental processes such as learning, analysing and memory, as well as important sensory faculties such as visual and audio abilities. In the mammals of lower orders this part of the brain can be well developed but is usually smooth on the surface; in mammals of a higher order, such as man and the cat, the surface area is increased by numerous convolutions and wrinkles which expand the brain's material capacity and make it more powerful.

While the cat's brain may not be as large as that of other

animals, the weight of its brain relative to the weight of its body is certainly greater. The average domestic cat today will have a brain to body-weight ratio of 1:99. A dog will typically have a ratio of 1:235 and a horse 1:593.

The weight of the brain is less important in determining intelligence than factors such as the amount of white and grey matter the brain contains, as well as the complexity of its structure. There is no doubt, however, that the brain of the cat is well developed and perfectly capable of processing complicated thoughts, besides having a range of sensory perceptions wider than our own.

THE INTELLIGENT CAT'S SUPERIOR APPEARANCE

To look wise is quite as good as understanding a thing, and very much easier.

· Oscar Wilde, *The Picture of Dorian Gray* ·

Even if cats did not have well-developed brains and excellent sensory abilities to facilitate intelligent thought, their appearance and attitude would certainly suggest the presence of an active mind. Cats are among the most graceful of creatures when they move, exuding an air of complete self-confidence and a command over their bodies and their environment. Even the least intelligent cats are almost always agile. The natural flexibility and control exhibited in their physical movements greatly surpass those of man.

The grace and co-ordination of its movements are only equalled, in the human sphere, by dancers.

· Patricia Dale Green, *Cult of the Cat* ·

The cat's body is much more flexible and limber than man's, with a skeleton of 244 bones compared with man's skeleton of 204 bones. Cats also have better muscular coordination; they are able to jump on to objects many times taller than their height and land on their feet after a fall.

One of the reasons that cats walk so elegantly and quietly is that they are all 'digitigrade'. In other words, they walk on their toes rather than on the soles of their feet. This not only enhances their walking style, but it is also very useful when stalking prey. Cats appear poised at all times too, since they are able to use their tails for balancing.

To complement their intelligent appearance, most cats maintain, with apparently very little effort, a convincing air of superiority. They often act in a lordly and aloof way towards those around them, either feigning or expressing a genuine feeling of disdain and a complete lack of interest.

To the intelligent cat, both its own activities and those of other people or animals can only captivate it for a short time, if at all. However, dull-witted cats can also get bored easily because they find it difficult to concentrate.

Many cats also have an endearing habit of perusing their surroundings like wise old men who are not only observing but philosophizing about everything they see. This happens most often when the cat is seated at a good vantage point, with its eyes slightly shut and head held high.

———

O little cat with yellow eyes,
Enthroned upon my garden gate,
Remote, impassive and sedate
And so unutterably wise.

Helen Vaughan Williams,
· *O Little Cat with Yellow Eyes* ·

———

The fact that cats are generally silent creatures contributes to their wise and knowledgeable image. It is easy to assume that their silence indicates a comprehensive understanding of the things around them. Given that in most aspects of their appearance and behaviour they seem intelligent, it is perhaps natural to imagine that their silent thoughts are intelligent as well.

THE INTELLIGENT CAT IN MYTH AND LEGEND

Long before it was ever proved in scientific experiment that cats are intelligent, it was widely assumed that they were, given their proud appearance, self-confident attitudes and extra-ordinary sensory abilities. A rich history of myths and legends is based on the belief that cats are not only intelligent creatures but that they also have a wide range of supernatural powers.

One of the most interesting and persevering themes among these legends is the belief that cats are clairvoyant. From the time of Ancient Egypt man has had the idea that the cat is capable of second-sight. It was appropriate that the Ancient Egyptian word for cat was 'mau' which means 'to see'.

Egyptians held the cat in high esteem and chose it as the animal representative for one of their most important goddesses, Bastet, who naturally shared attributes of fertility and second-sight. Bastet's sister goddess, Sekhmet, was also represented as part cat and correspondingly attributed with the power of second-sight. Because Sekhmet stood for justice and strength in war, a priest or priestess of the goddess would accompany soldiers to battle and draw upon Sekhmet's clairvoyant powers to predict where the enemy would strike.

Tomb paintings, statues and other depictions of Bastet often included an amulet carved or drawn in the shape of the Third Eye, the source of clairvoyant power. Symbolic of both Bastet and the cat, this sacred eye was known as the 'utchat' – possessed of mental and physical health.

Amulets of cats incorporating this image of the Third Eye were made in every shape, size and material imaginable and were kept in temples and homes or worn as jewellery for the seeing powers they were thought to bestow. If an amulet contained representations of both the cat and the eye, it was considered especially powerful, able to repel sickness and evil, and protect the souls of the dead.

The notion of the cat possessing the power of a Third Eye continued to thrive in the imagination of man after the decline of the Ancient Egyptian civilization. In Europe during the Middle Ages someone who was delirious and fevered would have a live cat passed over his or her body in order to 'see' clearly again. Second-sight was thought to be attainable by wearing a special charm made of the eyes of a black cat and the gall bladder of a human. In England at this time, children who played with tortoiseshell cats were believed to develop clairvoyance and were often encouraged to do so. And to recognize the Devil, the Jewish faith once taught its followers to burn the placenta from a black cat and rub the powdered ashes into one's eyes.

Another popular mythical theme related to clairvoyance is the belief that cats have extra-sensory perception (ESP). The cat's exceptional sensory abilities surpass those of man in almost every way but are not necessarily supernatural. Their superior eyesight and hearing do allow them to anticipate changes in the weather, earthquakes or just someone's impending arrival, and it is their 'early' reaction to such events that makes them seem to have ESP.

However, this does not explain how cats are able to 'sense' their way home, often from a great distance, after they have been lost or abandoned. Indeed, there is no scientific explanation of this phenomenon and whether it is due to telepathic ability or not is a matter for conjecture. None the less it does serve to perpetuate the myth that cats have extra-sensory perception.

Other legends assumed that the cat understood the workings of the heavens, the sun and the moon. The Ancient Egyptians believed that the eyes of the cat functioned in the same way as the moon at night, reflecting stored light from the sun, and, in another version of the myth, that they reflected light from the sun god Re, to which only cats and the moon goddess were privy. Like the moon and the sun, and the gods they represented, cats were considered capable of generating light to see in the dark.

The Chinese also believed that the cat was associated with the sun and would refer to cats' eyes to tell the time of day. Because the size of the pupil in a cat's eye alters according to the amount of light it receives, the eye can appear large and bright with the pupil in a narrow slit, or dark and brooding with the pupil fully expanded. So the eyelids of the cat would be lifted to determine how far the sun was above the horizon and therefore the approximate time of day. Elsewhere, it was believed that the current state of tidal flows could be discerned from the pupil in a cat's eye. A pupil that was fully expanded indicated an ebbing tide, whereas a contracted pupil in a thin slit signalled a tidal flood.

Cats also feature in the Chinese definition of the astrological calendar. Time is divided into twelve-year cycles with a different animal representing each of the twelve years. When people are born, they are assumed to embody many of the characteristics of the animal of that year. Those born in a Year of the Cat are considered to be clever, altruistic, discreet and refined, as well as devious, aloof and oversensitive. The last Year of the Cat was 1987.

Many myths credited the cat with an ability to predict the weather. The Scots believed that when a cat scratched at the leg of a table, gale-force winds were on their way. A myth of unknown origin claims that a cat seen washing behind its ears was an indication of rain. This idea was probably developed in a rainy country as most cats do this every day! The Chinese interpreted the wink of a cat's eye as a sign that it was about to rain. Even the whiskers of a cat were monitored to forecast weather: rain was imminent if they drooped, good weather was likely if they stood out straight.

In some parts of the world, a cat's sleeping position was thought to indicate certain weather patterns. If it slept with its head tucked down in between its paws, it was likely to rain. If it had its paws tucked underneath its body, it was about to turn cold. If it stretched out to its fullest length, warm weather was indicated. A cat seen sleeping with its front paws over its nose warned of gusty winds.

Some legends went on to presume that cats can influence the weather, and that they are privy to a wide range of supernatural powers which they share with the gods. In several cultures, for example, cats were carried ceremoniously around a dry field and then dipped into a pool of water to bring rain. Western Europeans believed that a cat could ensure a good harvest if one was buried alive in a freshly planted field. As representatives of the Ancient Egyptian sun god, Re, cats were thought to influence the sun and hence the fertility of the land, and to enhance the fertility of humans.

The Celts believed that cats were connected to the supernatural world and that strange fairies could be seen behind the eyes of a cat if one peered into them long enough. Similarly, they believed that fairies would use the eyes of cats to monitor what humans did. Perhaps this was because of the uncomfortable and hypnotic effect that cats' eyes can have on one who gazes into them.

In a book from 1584 entitled *Beware of the Cat* people were warned of the woeful consequences which were bound to befall the taker of a cat's life. The author believed that the cat would come back to exact its revenge, haunting such a person endlessly.

Conversely, in China, it was thought that cats could be used to *protect* man from evil spirits. Since the eyes of a cat could shine in the dark, they were believed to have special properties which included the ability to repel demons. At night many households would keep their cat chained outside so that it could guard the house from such spirits.

A long tradition of the cat's mythical healing powers also exists. The Roman historian Pliny believed that the excrement from cats could be mixed with various elements to create medically beneficial potions for humans. If mixed with oil of roses and resin, for instance, it was thought to eliminate uterine ulcers and if added to mustard it could cure ulcers which developed on the head. A heavy paste which included dung and a small amount of wine was recommended for removing thorns.

Edmund Topsell, the seventeenth-century natural historian, included a host of ways in which cats could be used for medicinal purposes in his book, *Historie of Foure-Footed Beastes and Serpentes*. He described at great length a method of curing blindness which required burning the head of a black cat until it turned into ashes and then applying the powder to the eyes using the quill of a feather. Other cures included the powder of dried cat's liver for stones in the bladder, and the extraction of fat from a cat was said to cure gout.

Cat skin and cat fur were also considered beneficial to the health of humans in many parts of the world. In Japan, for example, both epilepsy and gastritis were once treated by placing the fur of a black cat on top of the stomach. The skins of cats that had recently died were used in Holland and some other European countries to treat ailments such as skin diseases and throat infections. After the Great Fire of 1666 in London, burns were commonly covered with the soft fur from cat skins to protect the burnt area from the air, and so alleviate some of the pain.

The part of the cat considered most potent and used most often for the purposes of healing, though, was the tail. According to Celtic tradition if someone stepped on to the tail of a cat they were liable to be struck dead by a serpent. Throughout England the tails of black cats were often rubbed in a human's eye to eliminate styes. Inhabitants of Northamptonshire developed their own version of this myth, claiming that a single hair from a black tomcat's tail was enough to cure a stye if it was stroked with the hair nine times. As a general protection against ill health in the family, another myth called for the tail of a black cat to be buried under the doorstep of the house.

In severe cases of certain diseases, elaborate methods of treatment, which included the participation of a cat, were invoked. If someone was suffering from unbearable itching, writes Angela Sayer, 'a black cat was whirled by a left-handed man, three times around his head, then three drops of blood were let from its tail, to be mixed with the ashes of nine baked barleycorns. This potion was applied to the affected area with a gold wedding ring, while the Trinity was invoked and, hopefully, the irritation ceased.'

Today the concept of cat intelligence is much more object-ive. It is quantified and tested in laboratory experiments for scientific reasons, as well as for the improved understanding of our cats, so dispelling many superstitions and myths regarding the cat's power and abilities.

Yet, whether it is in the context of popular myth or scientific fact, the cat has almost always been considered by man as an intelligent creature, whose range of judgements and reasoning we may never fully understand. Even those who are not very clever can often make us believe they are by their appearance and the attitudes they adopt. But in most cases the cat's well-developed brain and superb sensory abilities make it as fascinating to watch as an animal as it is rewarding to keep as a pet.

III
THE DEFINITIVE
CAT I.Q. TEST

Cats are a mysterious kind of folk.
There is more passing in their minds
than we are aware of.

· Sir Walter Scott ·

———————

THE FOLLOWING set of questions is designed to test your cat's I.Q. in a way which is accurate but, above all, entertaining. The test consists of seventy-five questions which have been divided into four areas to analyse your cat's visual skills, audio abilities, social behaviour and domestic behaviour.

You should select only one answer for each question, responding as accurately as possible based on your observation and past experience of your cat. If a question refers to an unknown situation, try to imagine how your cat would probably react and respond accordingly. Similarly, if none of the possible answers to a question is applicable to your cat, choose the one which best approximates the response you would like to give. As you respond to the questions, mark each answer for your later reference to the scoring table which follows the test.

After you total the points for your cat, you can convert your score into a Cat I.Q., using the graph in the Results Analysis section. To see how your cat's I.Q. ranks among that of the domestic cat population, a detailed percentile ranking table is also included.

This domestic cat population is represented by a sample group of over sixty-five cats. In the SAMPLE POOL OF DOMESTIC CATS: LIST OF PARTICIPANTS at the back of the book, the name, breed and sex of each cat in the sample group is given.

HAVE FUN AND GOOD LUCK!

Part I Visual Skills

1. When it sits in a window, your cat:
   A Stares ahead blankly. ☐
   B Peruses what is outside with only general interest. ☐
   C Usually spots something outside and takes a strong interest in it. ☑

2. If your cat was to see a mouse running past, it would:
   A Capture it on its first attempt, then torment it. ☐
   B Certainly chase it but probably not catch it. ☑
   C Watch it run past. ☐
   D Attack it only if it was running near by. ☐

> *. . . take a cat, nourish it well with milk,*
> *And tender meat, make it a couch of silk,*
> *But let it see a mouse along the wall,*
> *And it abandons milk and meat and all,*
> *And every other dainty in the house,*
> *Such is its appetite to eat a mouse.*
>
> · Chaucer, *Canterbury Tales* ·

3. Gather up your coat and keys, as if you were about to leave the house. Your cat:
   A Accompanies me to the door affectionately, saddened that I am leaving. ☑
   B Moves to the kitchen, ready to plunder it once I go. ☐
   C Miaows as if to say goodbye. ☐
   D Does not take any notice at all. ☐

4. Does your cat hide objects around the house and, if so, does it hide them well?

A I am sure it does, but I have never discovered where. ☐

B Yes, it does and I occasionally discover them later. ☐

C No, it does not seem to occur to my cat to hide anything. ☐

D I don't really know. ▨

5. Take out an unopened tin of cat food and place it in front of your cat. Your cat:

A Investigates the tin as it would any other item, losing interest quickly. ☐

B Goes to wait at its food bowl, assuming I am about to feed it. ▨

C Recognizes the tin and gets excited but can't understand what I am doing. ☐

D Ignores my actions altogether. ☐

6. If you move house or just rearrange the furniture, your cat:

A Cautiously explores every corner of the room(s), taking its time to feel comfortable. ▨

B Finds a secure hiding-place and remains there for days. ☐

C Miaows and is generally disoriented for some time. ☐

D Quickly and confidently settles into the new environment. ☐

7. If your cat was running towards a low-lying table or other piece of furniture under which it wanted to hide, how well would it gauge the amount of headroom it has?

   A Poorly. It would not slow down on its approach and would probably hit its head. ☐

   B Very well, but it would go under the furniture in a slow and careful manner. ☐

   C Expertly. I've seen my cat do this many times. ▨

8. Take a morsel of meat – or any other food your cat likes – and dangle it in front of its nose. Then hide the food behind your back. Your cat:

   A Looks at me with disdain and walks off. ☐

   B Exhibits little interest because it doesn't seem to realize that the food is there, out of sight. ☐

   C Tries to get behind my back to reach the food. ▨

   D Looks confused and unhappy, so I give it the food after a while. ☐

9. The door to your flat or house has been left open and your cat notices. Does your cat:

   A Nudge the door shut with its paw. ☐

   B Dismiss the open door altogether. ☐

   C Take the opportunity to go outside to explore. ▨

   D Tentatively approach the open door, but not leave the doorstep. ☐

10. Stare your cat in the face and when it looks back at you, smile at it. Your cat:

    A Looks at me as if I've lost my mind.

    B Runs away.

    C Doesn't react at all.

    D Approaches me, perhaps starting to purr.

11. Cats can see in light over five times dimmer than man can. How would you rate your cat's night vision abilities?

    A Below average. It is sometimes clumsy and noisy when it moves around at night.

    B I don't know because it spends the entire night sleeping.

    C Excellent. It's more active at night than during the day.

    D Good. It moves about the house confidently and quietly at night.

12. How adept is your cat at judging distance when jumping on to a windowsill or table?

    A Very skilled. It never misses.

    B Usually very good, although it has incorrectly estimated heights a few times.

    C I don't know because my cat does not bother to jump up on to surfaces.

13. Your cat is facing the television screen when a
commercial featuring cats comes on. Your cat:

   A Watches the action with no especial interest. ☐

   B Recognizes a 'cat' and may extend a paw towards ☒
the image.

   C Simply stares blankly at the screen. ☐

   D Recognizes a 'cat' and miaows or hisses in fright. ☐

14. Your cat is facing the television screen when the
picture shows birds in flight. Does your cat:

   A Continue to face the television, but not react to ☐
what's on the screen.

   B Proceed to fall asleep. ☐

   C Watch the birds fly across the screen, and then ☒
look beyond and behind the television to discover
where they went.

   D Try to 'attack' the birds. ☐

39

15. Put a small portion of your cat's food into its bowl and just after it has started eating, interrupt it by putting the food into a plastic bag. Close the bag and place it next to your cat's bowl. Your cat:

A Is obviously irritated and annoyed at this disturbance, but wastes no time as it rips open the bag to continue eating. ☐ !

B Thinks I am playing a game and happily tries to open the bag. ☐

C Investigates and smells both the bag and the empty bowl, but does not try to open the bag. ☐

D Is not sure what has happened. ☐

———————

*If we take the case of cats, they say little,
but they think a great deal; they conduct trains of reasoning.*

· Andrew Lang (1844–1912) *Longman's Magazine* ·

———————

Part II Audio Abilities

16. If you are in the kitchen and open the refrigerator door, your cat:

    A Would suddenly come into the kitchen. ☐

    B Might come in if it already happened to be near by. ☐ /

    C Would probably walk past the kitchen. ☐

17. With your cat near by, start to sing or hum a tune and watch your cat's reaction. Your cat:

    A Expresses its low opinion of my musical ability by leaving the room. ☐

    B Is oblivious and continues whatever activity it was engaged in before the singing began. ☐

    C Goes under a chair for cover. ☐

    D Listens attentively. ☐

18. Your cat is sleeping peacefully when suddenly there is a loud and unusual noise from the opposite side of the house. Your cat:

    A Would not move at all. ☐

    B Would spring into action and immediately investigate the sound. ☐

    C Might stroll to the scene, but would appear annoyed at the disturbance. ☐

19. Does your cat realize when it is being talked about?

    A Yes. ☐

    B Some of the time and it enjoys the attention. ☐

    C No. It ignores all conversations. ☐

20. When you call for your cat, using its name, it:

   A Responds and invariably comes to me. ☐

   B Recognizes the call but may or may not act on it. ☐

   C Appears not to register its name at all. ☐

21. When others call your cat by its name, your cat:

   A Obviously recognizes that its name has been called, but doesn't respond since it is not its owner's voice. ☐

   B Doesn't even seem to notice there are people around. ☐

   C Occasionally recognizes its name and comes over when it does. ☐

   D Only responds if the caller has food to offer. ☐

22. How often does your cat respond to the sound of food being placed in its bowl?

   A Sometimes, if it is hungry. ☐

   B Never. I often have to remind it where the food bowl is. ☐

   C Always, as if it had not eaten in days. ☐

23. How often does your cat sit by the door and listen to outside noises?

   A The concept of an 'outside' is still foreign to my cat. ☐

   B Frequently, often with its nose or paw actually poking beneath the door. ☐

   C Perhaps once or twice a day, but in a rather listless fashion. ☐

   D Never. ☐

24. A loud thunderstorm rolls in during the night. Your cat would:

    A Be happy just to be inside. ☐

    B Panic and miaow as if the world were about to end. ☐

    C Act scared as an excuse to sleep on my bed. ☐

    D Take no notice. ☐

25. When you are speaking to someone on the telephone, does your cat:

    A Think I'm talking to it. ☐

    B Start playing with the cord or the phone itself. ☐

    C Recognize that I'm not speaking to it, and stares at me as if to say 'Please stop talking!' ☐

    D Not respond at all to my conversation, regardless of the volume or tone of my voice. ☐

26. The television is on and shows kittens crying or birds singing. Your cat:

   A Begins to miaow, wondering where the noise is coming from. ☐

   B Determines that the sound comes from the television, but continues to look perplexed. ☐

   C Pays no attention to the sound. ☐

   D Understands that the sound corresponds to the image and looks frantically behind the television for a 'way in'. ☐

27. Take your hairdrier or any other noisy electrical appliance and plug it in. Make sure your cat is watching before you switch it on. Your cat:

   A Calmly leaves the room as soon as it sees the appliance. ☐

   B Quietly disappears after I've turned it on. ☐

   C Bolts from the room as soon as I switch it on. ☐

   D Stays put, quite oblivious to the noise. ☐

28. If you are in the kitchen and begin to unwrap food, your cat:

   A Takes no interest in it whatsoever. ☐

   B May take a while to respond, but is invariably curious. ☐

   C Will hear immediately and rush over for some. ☐

   D Will wait for me to give it its share. ☐

29. If your cat is near a door or window and hears a strange noise outside, it would probably:

   A Try to get through the door or window to determine what the noise is. ☐

   B Stay where it is, but continue to monitor the noise. ☐

   C Tentatively get closer to the door or window. ☐

   D Get scared and run away to hide. ☐

30. If you call your cat, using a word other than its name but in the same intonation as your usual call, your cat:

   A Would come over as if its name had been called. ☐

   B Would not respond at all. ☐

   C Would recognize the tone, but not respond because its name had not been used. ☐

   D Might surprise me and come over, though this never happens when I call it by its name. ☐

Part III *Social Behaviour*

The cat lives alone. He has no need of society.
He obeys when he wishes, he pretends to sleep the better to see,
and scratches everything he can scratch.

· Vicomte de Chateaubriand (1768–1848) ·

---

31. When it meets a dog, your cat:

    A Greets it as it does any other animal, without hesitation. ☐

    B Stays out of the way as dogs are bigger than cats. ☐

    C Ignores it as cats are far superior to dogs. ☐

    D Triggers the dog's thoughts on lunch. ☐

32. When it meets another cat, your cat usually:

   A Acts aloof. ☐

   B Picks a fight. ☐

   C Is petrified. ☐

   D Tries to play with it. ☐

No matter how much cats fight,
there always seem to be plenty of kittens.

· Abraham Lincoln (1809–65) ·

---

33. The feline vocabulary is diverse, including all varieties of miaows, hisses, purrs and growls. How vocal is your cat?

   A Extremely. I wish it would be quiet more often. ☐

   B Quite communicative. ☐

   C Not very vocal at all. ☐

   D Distinctly unvocal. Sometimes when it miaows, no sound comes out. ☐

34. How well does your cat interpret cat vocabulary used by yourself, another person or on television?

   A It pays little or no attention to the sounds made by others. ☐

   B It usually seems to understand. ☐

   C Expertly. ☐

   D It only understands if I make the sounds. ☐

35. When transporting your cat in a pet carrier, your cat tends to:

A Act unsettled, although it keeps its composure. ☐

B Enjoy the trip. ☐

C Tear around the carrier, panic-stricken and yowling. ☐

D Fall asleep. ☐

36. If your cat were to throw a dinner party, which of the following foods would it be most likely to serve?

A Smoked salmon. ☐

B Go Cat or any other cat food. ☐

C Caviar, served with blinis, fresh cream and a squeeze of lemon. ☐

D Canned tuna. ☐

How did cats acquire their taste for fish?
Has anyone ever seen a wild cat angling?

Andrew Lang (1844–1912),
· *Longman's Magazine* ·

---

37. When a stranger visits your house, your cat:

A Acts defensively or in a hostile way. ☐

B Greets him or her enthusiastically, hoping to be stroked. ☐

C Doesn't give the person a second look. ☐

D Will become friendly if the stranger starts eating. ☐

38. Your cat has just accomplished something, such as catching an insect or finding some food you had put away. Your cat:

A Assumes its 'I'm the best there is' look. ☐

B Does not change its expression as it always has its 'I'm the best there is' look. ☐

C Is ready to move on to the next challenge. ☐

D Looks surprised but pleased at its achievement. ☐

39. How strong is your cat's natural instinct to fight?

A If it has one, it is entirely dormant. It seems to think that fighting is a waste of time that could be spent sleeping or eating. ☐

B My cat loves to fight and hates to lose. It turns the simplest game into a confrontation or contest. ☐

C My cat's fighting instinct is strong, but only when called for. ☐

D My cat only likes to fight in a playful way. ☐

There were once two cats of Kilkenny.
Each thought this was one cat too many.
So they fought and they fit
And they scratched and they bit,
'Til excepting their nails
And the tips of their tails
Instead of two cats there weren't any!

· Traditional limerick ·

---

40. How does your cat behave with children?

   A In the same aloof way it behaves with adults. ☐

   B Apprehensively, as it knows that children can get rough when they play. ☐

   C Relaxed and delighted if they want to play. ☐

   D Annoyed by their antics, especially if they steal the limelight. ☐

41. What would your cat think if you got a new kitten?

   A 'I'm sure I was never that immature.' ☐

   B 'Why is it so *small*?' ☐

   C 'A new playmate!' ☐

   D 'When are we getting rid of this nuisance?' ☐

42. How difficult is it to please your cat?

   A Not difficult at all; it is very easy-going. ☐

   B It is hard to please when it's upset about something. ☐

   C Quite difficult. ☐

   D So difficult that I wonder if it *enjoys* its state of perpetual dissatisfaction. ☐

43. If your cat decided it would like a drink but there is nothing in its bowl, it would:

   A Miaow sweetly and tap at the bowl with its paw to give me the message. ☐

   B Miaow loudly by its bowl to attract my attention. ☐

   C Hope that I will notice the bowl is empty and wait patiently. ☐

   D Try to sip from my glass, if there's something in it. ☐

*Long contact with the human race has developed in
[the cat] the art of diplomacy, and no Roman Catholic
of medieval days knew better how to ingratiate
himself with his surroundings than a cat with a
saucer of cream on its mental horizon.*

· Saki, *The Achievement of the Cat* ·

44. If your cat could read, which of the following
newspapers would it probably buy?

A *Financial Times.* ☐

B *Daily Mail.* ☐

C *Independent.* ☐

D *Sun.* ☐

45. If your cat was fed next to another cat, it would:

A Ignore it and eat its own food. ☐

B Barge over to the other cat's bowl and eat as
much as it could. ☐

C Not only hijack the other cat's bowl but keep an
eye on its own bowl so that the cat doesn't eat
its food. ☐

D Refuse to eat at all with another cat present. ☐

*. . . many cats, and one of mine in particular,
always desert their own platter (however tempting)
for that of their neighbour.*

· Andrew Lang (1844–1912), *Longman's Magazine* ·

46. Take out your pet carrier and place it in front of your cat. It will:

A Grudgingly get in when asked. ☐

B Run and hide right away. ☐

C Not react because it knows it is being tested. ☐

D Jump right in without any prompting. ☐

47. When you and your cat are in the vet's waiting-room, how does your cat behave?

A Very apprehensively. The mere mention of the V word is enough to send it running. ☐

B Frightened and worried, but only when it actually recognizes the veterinary surgery. ☐

C Confused and disgruntled, but not scared. ☐

D Aggressively, ready to attack when the vet appears. ☐

48. Does your cat ever appear to remember people who visit only occasionally?

A No, not at all. ☐

B Sometimes, especially if the person likes cats and paid attention to it. ☐

C Yes, if the friend gave it food on his/her last visit. ☐

D No, but it will pretend to remember if the acquaintance offers it food on *this* visit. ☐

49. Assuming that your cat was one of a litter, how do you think it would behave towards its brothers and sisters if they were present now?

A It would mother them. ☐

B In a high-handed manner with an air of utter superiority. ☐

C It would play with them every now and then. ☐

D It would not pay much attention to them. ☐

50. How does your cat react when you bring out the camera?

A It strikes a pose. ☐

B Completely uninterested, unless I use the flash and then it bolts out of the room. ☐

C It starts to inspect the camera and won't stay still. ☐

D It continues whatever it's doing, as it's not at all bothered by the camera. ☐

Cats make exquisite photographs.
They don't keep bouncing at you to be kissed
just as you get the lens adjusted.

Gladys Taber,
· *Ladies Home Journal*, October 1941 ·

---

51. How do you think your cat would spend its free time if it was human?

A Sleeping. ☐

B Reading books and/or playing with computers. ☐

C Watching television. ☐

D Cooking and/or throwing parties. ☐

52. In general, how affectionate is your cat?
    A So affectionate that it doesn't like to leave my side. ☐
    B Quite affectionate. ☐
    C Decidedly unaffectionate. It is too busy sleeping or being aloof. ☐
    D When it comes to affection my cat is an opportunist, turning on the charm only when it wants something. ☐

53. When you read your favourite magazine, your cat:
    A Occasionally looks at it with me. ☐
    B Jumps up to sleep beside me. ☐
    C Jumps up and blocks my view until I push it off. ☐
    D Continues to jump up and block my view no matter how many times I push it off. ☐

Part IV Domestic Behaviour

---

54. The average cat sleeps eighteen hours a day.
In general, how many hours does your cat sleep?

A All day, if possible. ☐

B No more than eighteen hours. ☐

C About ten hours. ☐

D Five hours or less. ☐

---

*Cats are rather delicate creatures,
and they are subject to a good many ailments,
but I never heard of one who suffered from
insomnia.*

· Joseph W. Krutch (1893–1970) ·

---

55. Of the following types of food, which would your
cat probably choose as its favourite?

A Any food from my plate. ☐

B Any unusual or expensive food. ☐

C Almost any food, even inedible items such as
socks or newspapers. ☐

56. If you bounce a ball in front of your cat, it will:

A Play until the ball rolls out of easy reach. ☐

B Aggressively chase the ball. ☐

C Play happily, sometimes for hours. ☐

D Show no interest in it. ☐

57. How does your cat let you know when it wants to go outside or get through a closed door?

A It scratches gently at the door. ☐

B It sits at the door and miaows continuously. ☐

C It rarely needs to as it can open most doors by itself. ☐

D It sits silently facing the shut door waiting for it to open miraculously. ☐

58. There is a fly in the house. Your cat:

A Appears uninterested, but is ready to attack when the fly gets within striking distance. ☐

B Ignores it altogether. ☐

C Chases it incessantly. ☐

D Would ask 'What is a fly?' ☐

What intenseness of desire
In her upward eye of fire!
With a tiger-leap half-way
Now she meets the coming prey . . .

William Wordsworth,
· 'The Kitten and Falling Leaves' ·

59. If your cat was by itself in another part of the house and wanted to find you, it would probably:

A Wander around the house and miaow occasionally until it found me. ☐

B Miaow incessantly until I appeared. ☐

C Listen attentively for any noises I might make to indicate where I was and then head in that direction. ☐

D Wonder where I was but simply wait for me to reappear. ☐

60. At teatime or any mealtime, your cat:

A Takes a seat at the table five minutes before I do. ☐

B Joins me when it hears the clatter of plates, then sits under my chair. ☐

C Always happens to be there at that time and decides to loiter. ☐

D May or may not come in. ☐

———

When the tea is brought at five o'clock,
And all the neat curtains are drawn with care,
The little black cat with bright green eyes
Is suddenly purring there.

At first she pretends, having nothing to do,
She has come in merely to blink by the grate.
But, though tea may be late and the milk may be sour,
She is never late.

· Harold Monro, *Milk for the Cat* ·

———

61. How often does your cat follow you around the house?

A Occasionally, but no more than 30 per cent of the time. ☐

B Almost always, even following me outside. ☐

C About 50 per cent of the time. ☐

D Never, unless it happens to be going the same way. ☐

62. How curious is your cat?

A Incredibly. It spends most of its time investigating every object, activity and noise both inside and outside the house. ☐

B Fairly curious, but investigating is not its favourite pastime. ☐

C The only thing it is ever curious about is what's for dinner. ☐

---

*Curiosity killed the cat,
satisfaction brought it back.*

· Wise old saying ·

---

63. What does your cat like to do with curtain cords and other hanging items in your home?

A Attacks them. ☐

B Tosses them back and forth. ☐

C Ties them into knots. ☐

D Ignores them. ☐

64. It is said that cats have nine lives. Assuming that your cat is on life number one at the moment, how do you think it would spend its next eight lives?

A Sleeping. ☐

B Sleeping and eating. ☐

C Sleeping often, but inspecting its surroundings daily and playing lots of games. ☐

D Doing its best to be a real nuisance without being admonished. ☐

65. Is your cat a prankster?

  A It could be but knows better. ☐

  B Only rarely, when it is in a spirited mood. ☐

  C No. Not because it doesn't *want* to play tricks,
but because it probably can't think of any to play. ☐

  D Yes. It seems to live for such antics. ☐

66. If you ask your cat to come with you to another
room, or even go outside, how often will it obey
you?

  A I'm not sure. I suspect that when it does, it's just
by chance. ☐

  B Hardly ever. ☐

  C Almost always. ☐

  D Never. Even if it was actually on its way there, it
would go elsewhere so as not to look subservient. ☐

67. What is your cat's favourite sleeping position?

A Snuggled up in a ball. ☐

B Squeezed inside the smallest container it can fit into. ☐

C Stretched out on its back. ☐

D Any position, as long as it's somewhere on top of me. ☐

68. If your cat was to climb a very tall tree, how skilled would it be at getting down?

A Reasonably adept, but it would take a very long time to get down, hesitating at every step. ☐

B Very adept. It could climb down almost as quickly as it climbed up. ☐

C Quite adept. It would climb down in a cautious manner. ☐

D It would be completely helpless and have to be rescued. ☐

A cat can climb down from a tree without the assistance of the fire department or any other agency.
The proof is that no one has ever seen a cat skeleton in a tree.

· Anonymous ·

69. When you play with your cat, how long on average is its attention span?

A Less than five seconds. ☐

B Up to ten seconds. ☐

C Around thirty seconds. ☐

D Until it believes it has won the game, or when I give up. ☐

70. At night, does your cat:

    A Get up occasionally, usually for a drink or a snack. ☐

    B Continually change its position on my bed, choosing the warmest spot. ☐

    C Sleep right through, exhausted after a day spent sleeping. ☐

    D Stay awake for most of the night, as if on night duty. ☐

71. How strong is your cat's desire to scratch objects and sharpen its claws?

    A It only occasionally scratches. ☐

    B It scratches solely for the thrill of destruction. ☐

    C It has a healthy interest and uses a scratching post or other object for scratching. ☐

    D Non-existent. Scratching does not interest my cat. ☐

72. Does your cat let you know when it's ready for its next meal and, if so, how does it convey the message?

A By miaowing near its food bowl.

B By rubbing against my legs and miaowing sweetly.

C My cat just wanders over to its food bowl when it is hungry, hoping to find food there.

D By miaowing in a demanding tone and giving me a hard stare.

Cats seem to go on the principle that it never does any harm to ask for what you want.

· Joseph W. Krutch (1893–1970) ·

---

73. How often does your cat groom itself?

A Once or twice a day.

B Three to five times a day.

C Up to ten times a day.

D Many times throughout the day. It's my cat's favourite activity after sleeping and eating.

74. Since you've had your cat, have you noticed that occasionally your things go missing?

A Yes, from food and jewellery to socks and pens.

B No, it never touches any belongings.

C A few items, perhaps.

D Yes, several items have gone missing, but they always resurface later.

A dog will often steal a bone,
But conscience lets him not alone,
And by his tail his guilt is known.

But cats consider theft a game,
And, howsoever you may blame,
Refuse the slightest sign of shame.

When food mysteriously goes,
The chances are that Pussy knows
More than she leads you to suppose.

And hence there is no need for you,
If Puss declines a meal or two,
To feel her pulse and make ado.

· Anonymous ·

75. How would your cat react if you tried to give it a bath?

A Insulted that I should consider its appearance to be less than perfect.

B Agitated and annoyed that it didn't see it coming.

C It would suddenly vanish into thin air the moment I *thought* about giving it a bath.

D It would enjoy the bath and make sure that I got soaking wet, too.

SCORING TABLE

Part I Visual Skills

Question 1
A = 1
B = 2
C = 3

Question 2
A = 4
B = 3
C = 1
D = 2

Question 3
A = 2
B = 4
C = 3
D = 1

Question 4
A = 4
B = 3
C = 1
D = 2

Question 5
A = 2
B = 4
C = 3
D = 1

Question 6
A = 3
B = 1
C = 2
D = 4

Question 7
A = 1
B = 2
C = 3

Question 8
A = 3
B = 2
C = 4
D = 1

Question 9
A = 3
B = 1
C = 4
D = 2

Question 10
A = 4
B = 1
C = 2
D = 3

Question 11
A = 2
B = 1
C = 4
D = 3

Question 12
A = 3
B = 2
C = 1

Question 13
A = 2
B = 4
C = 1
D = 3

Question 14
A = 2
B = 1
C = 4
D = 3

Question 15
A = 4
B = 3
C = 2
D = 1

Part II Audio Abilities

Question 16	Question 17	Question 18	Question 19
A = 3	A = 4	A = 1	A = 3
B = 2	B = 1	B = 3	B = 2
C = 1	C = 2	C = 2	C = 1
	D = 3		

Question 20	Question 21	Question 22	Question 23
A = 2	A = 3	A = 2	A = 1
B = 3	B = 1	B = 1	B = 4
C = 1	C = 2	C = 3	C = 3
	D = 4		D = 2

Question 24	Question 25	Question 26	Question 27
A = 2	A = 3	A = 2	A = 4
B = 1	B = 2	B = 3	B = 3
C = 4	C = 4	C = 1	C = 2
D = 3	D = 1	D = 4	D = 1

Question 28	Question 29	Question 30
A = 1	A = 4	A = 2
B = 2	B = 3	B = 3
C = 3	C = 2	C = 4
D = 4	D = 1	D = 1

Part III Social Behaviour

Question 31	**Question 32**	**Question 33**	**Question 34**
A = 3	A = 4	A = 3	A = 1
B = 2	B = 3	B = 4	B = 2
C = 4	C = 1	C = 2	C = 4
D = 1	D = 2	D = 1	D = 3

Question 35	**Question 36**	**Question 37**	**Question 38**
A = 2	A = 3	A = 3	A = 2
B = 4	B = 1	B = 2	B = 4
C = 1	C = 4	C = 1	C = 3
D = 3	D = 2	D = 4	D = 1

Question 39	**Question 40**	**Question 41**	**Question 42**
A = 1	A = 4	A = 3	A = 1
B = 4	B = 1	B = 1	B = 2
C = 3	C = 2	C = 2	C = 3
D = 2	D = 3	D = 4	D = 4

Question 43	**Question 44**	**Question 45**	**Question 46**
A = 2	A = 4	A = 2	A = 2
B = 4	B = 2	B = 3	B = 3
C = 1	C = 3	C = 4	C = 4
D = 3	D = 1	D = 1	D = 1

Question 47	**Question 48**	**Question 49**	**Question 50**
A = 4	A = 1	A = 2	A = 4
B = 2	B = 2	B = 4	B = 1
C = 1	C = 3	C = 3	C = 3
D = 3	D = 4	D = 1	D = 2

Question 51	Question 52	Question 53
A = 1	A = 1	A = 4
B = 4	B = 2	B = 3
C = 2	C = 3	C = 2
D = 3	D = 4	D = 1

Part IV Domestic Behaviour

Question 54	Question 55	Question 56	Question 57
A = 1	A = 2	A = 1	A = 2
B = 2	B = 3	B = 4	B = 3
C = 3	C = 1	C = 2	C = 4
D = 4		D = 3	D = 1

Question 58	Question 59	Question 60	Question 61
A = 4	A = 3	A = 4	A = 3
B = 2	B = 2	B = 2	B = 1
C = 3	C = 4	C = 3	C = 2
D = 1	D = 1	D = 1	D = 4

Question 62	Question 63	Question 64	Question 65
A = 3	A = 3	A = 1	A = 2
B = 2	B = 2	B = 2	B = 3
C = 1	C = 4	C = 3	C = 1
	D = 1	D = 4	D = 4

Question 66	Question 67	Question 68	Question 69
A = 1	A = 2	A = 2	A = 1
B = 3	B = 1	B = 4	B = 2
C = 2	C = 3	C = 3	C = 3
D = 4	D = 4	D = 1	D = 4

Question 70	**Question 71**	**Question 72**	**Question 73**
A = 2	A = 2	A = 3	A = 1
B = 3	B = 4	B = 2	B = 2
C = 1	C = 3	C = 1	C = 3
D = 4	D = 1	D = 4	D = 4

Question 74	**Question 75**
A = 4	A = 2
B = 1	B = 3
C = 2	C = 4
D = 3	D = 1

RESULTS ANALYSIS
Cat I.Q. Conversion Graph

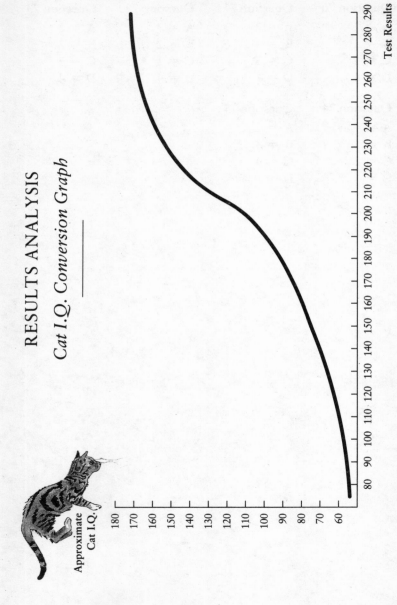

Approximate Cat I.Q.

Test Results

Cat I.Q. Percentile Ranking Table

CAT I.Q.	CAT I.Q. CATEGORY	PERCENTAGE OF CAT POPULATION MORE INTELLIGENT THAN YOUR OWN
79 and below	Blissfully Ignorant	97%
80–94	Occasionally Quite Clever	73%
95–105	Average	50%
106–120	Brighter than Average	35%
121–135	Very Intelligent	22%
136–155	Extremely Intelligent	8%
156 and above	Possibly Brighter than You Are	3% or Less

IMPORTANT MESSAGE

This test has been designed with a greater emphasis on entertainment than accuracy. Results are subject to wide variation and therefore should not be used as a basis for any personal decisions about your cat!

(*Source:* Sample Pool of Domestic Cats)

IV
MAN AND CAT:
THEIR RELATIONSHIP THROUGH
THE AGES

———

THE RELATIONSHIP between man and the domesticated cat has gone through different phases in history and has now almost come full circle. Cats today are much loved and well-respected members of the family in millions of households. They enjoy the best of modern creature comforts and in most cases have well-trained owners to tend to their needs and desires.

But life with man has not always been pleasant for cats. They were traded as commodities by the Ancient Greeks and Romans, who valued them primarily for their ability to catch rodents. They were persecuted in the Middle Ages as they were believed to be instruments of the Devil. Yet, at the beginning of their relationship with man, cats, as today, enjoyed a favourable status within the human community.

Over ten thousand years ago, man domesticated animals such as the dog, horse and cow, who sacrificed their freedom in exchange for the food and shelter that man could provide. There are no records, however, of domesticating cats at this time and it was apparently the cat who decided to live with man about five thousand years later in the communities of Ancient Egypt.

This popular assumption may be challenged, however, by a discovery made in Cyprus in 1983. During excavations at a

75

Neolithic settlement, Alain le Brun discovered the jawbone of a cat which was approximately eight thousand years old. The fact that it was found in Cyprus is significant because cats were not indigenous to the island and must have been brought there originally by settlers.

There is evidence that these settlers also brought other kinds of domesticated animals and it is difficult to imagine that they would have brought cats other than fully domesticated ones. 'A spitting, scratching, panic-stricken wild feline,' argues Desmond Morris in his book *Catlore*, 'would have been the last kind of boat-companion they would have wanted. Only tame, domesticated animals could possibly have been part of the goods and chattels of that early band of pioneers, striking out for a new island home.'

This piece of evidence stands on its own, however. The next known records of the domesticated cat date from about 2600 B.C. and are found in paintings and inscriptions from tombs and other ruins of the Ancient Egyptian civilization. It is thought that cats were attracted to human communities there because of the profusion of rodents that lived off the food stores and granaries which the Egyptians had begun to build. The first cats were probably of the *Felis lybica* variety, still in existence today. A short-haired and lean cat species from Libya and Egypt, the *Felis lybica* is the most likely ancestor of today's domestic cat. It has a long and elegant body with striped or spotted fur usually of a red or brown colour.

Although initially considered intruders in the community, cats were eventually accepted and later valued by the Egyptians for controlling disease-infested vermin, as well as for their charming and often intriguing personalities. Cats soon found their way into the households of Ancient Egypt and their standing in society continued to improve for several thousand years.

Cats came to be considered beloved members of the family

and were portrayed quite regularly on wall paintings and frescoes after 1600 B.C., playing and eating indoors, sitting underneath their owners' chairs, and engaging in other domestic activities.

But although the cat was eventually treated well in most homes, it was soon elevated to a much higher status in society when it came to represent the goddess of fertility, Bastet. Bastet was known also as Bast and Pasht, from which our word 'puss' is most likely derived. It was appropriate that Bastet and her sister Sekhmet should be represented as part feline since their father Re was usually depicted as part human and part cat. Indeed, it was believed that Re, as the sun god, would turn himself into a cat every day at dawn in order to kill the night's serpent of darkness, Apep.

The cat was especially well-placed as a symbol of fertility as it was able to have numerous kittens on a regular basis. Plutarch wrote '[the cat] brings forth at first one kitten, afterwards two, and the third time three; and that number increaseth thus until the seventh and last birth, so that she bears in all twenty-eight young, or as many as the moon hath revolutions'. In fact cats are even more fertile than Plutarch believed and can have over a hundred kittens during a lifetime if they are left to breed.

Cats were also preferred to lions as representatives of Bastet because they were easier to manage and less dangerous to worship and keep in a temple. In temples across Egypt cats were watched by priests twenty-four hours a day; their every movement, miaow and purr was interpreted as a message from the goddess.

Bastet's high standing was ensured in 1000 B.C. when a new line of pharaohs took over the throne in Egypt. Of Libyan origin, these pharaohs were ardent worshippers of Bastet. When they established their capital in Bubastis, known as Tell Basta today, the capital became the religious centre for cult worship of the goddess. A festival in Bastet's honour was held every year in the city and included a large fair, a spectacular procession and the consumption of more wine than at any other time of the year. Over seventy thousand people would attend the celebrations, many travelling for days on the Nile to get there. The Greek historian Herodotus described the festival in 450 B.C. as the most lavish in all Egypt.

The temple at Bubastis was noted by Herodotus as well: 'Here is a temple of Bubastis deserving of mention; other temples are larger and more magnificent but none more beautiful than this . . .' Constructed of red blocks of granite, it was built on a large square; canals one hundred feet wide, fed by the Nile, flowed down either side. A shrine to Bastet's glory was placed in the temple's centre, protected by tall trees encircled by a stone wall.

An inscription and painting on an interior wall of the temple is a testament to Bastet's importance. She is shown receiving gifts from King Osorkon the Second, who praises her with the following words: 'To thee I give every land in Obeisance. To thee I give all power like Re.'

Domestic cats were worshipped and portrayed in countless paintings, sculptures and engravings across Egypt. Young wives would hang amulets of cats on the walls of their home and pray to Bastet, as the goddess of fertility, that they would be blessed with children. Jewellers used the image of cats as a decorative motif and, in turn, cats were portrayed wearing jewellery themselves, usually necklaces or earrings made out of precious metals and stones.

Cats were considered so sacred that to kill one, even by accident, could mean punishment of death. In the case of a house fire, the cat was to be saved before any members of the family or household items.

In 500 B.C. their high regard for cats worked against the Egyptians when the Persians attacked the Egyptian town of Pelusium. As the Persians were losing the battle, the chief of their army, Cambyses, ordered his men to retreat and collect as many cats from the area as they could find. When they next attacked the town, each Persian soldier was carrying a cat, and a large group of cats also ran free just ahead of and around the army. This rendered the defending Egyptians helpless, as they would not risk killing even one cat, and the town promptly surrendered without any further combat.

Several hundred years later, in 50 B.C., the Sicilian historian Diodorus was amazed to discover that the Egyptians still held the cat in greater esteem than man or politics. He wrote the following account of an incident involving a cat in Egypt and the associated death of a visiting Roman:

> And because of their fear of such a punishment, any who have caught sight of one of these animals lying dead withdraw to a great distance and shout with lamentations and protestations that they found the animal already dead. So deeply implanted also in the heart of the common people is their superstitious regard for these animals ... that once ... when one of the Romans killed a cat and the multitude rushed in a crowd to his house, neither the officials sent by the king to beg the man off nor the fear of Rome which all the people felt were enough to save the man from punishment, even though his act had been an accident.

When a cat did die, those bereaved by the cat's death would go into long periods of mourning, traditionally shaving their eyebrows and beating on death gongs as expressions of their grief. Like humans, cats were assumed to have an afterlife and therefore they were usually mummified and given their own burial rights and funerals.

A deceased cat's body would be prepared and encased in linen – plain linen if the cat's owner was poor and multi-coloured if wealthy. A papier-mâché mask would be placed on the head of the cat with its eyes represented by circles of linen and its ears by ribbed palm leaves. Then the body would be placed in a coffin which might have a cat's face painted on the top of its case, with eyes of precious and semi-precious stones.

To sustain a cat in the afterlife, food, such as mummified mice or other small animals, was often included in the coffin or tomb. In one tomb dating from 1700 B.C., seventeen cat skeletons were discovered, each of which had been well provided for with its own small bowl for milk.

Cat cemeteries flourished across the country, with a large and well-known burial ground discovered at Bubastis itself. During excavations in the late nineteenth century of the ancient city of Beni-Hassan in central Egypt, over 300,000 mummified cat remains were found in a single cemetery.

The influence of cats extended to the rituals of funeral ceremonies for humans. A deceased human was often buried with a small ivory wand decorated with the head of a cat. This was intended to represent the cat who, according to a widely held belief at the time, would guide and protect the deceased's soul as it travelled to the underworld.

The Egyptians maintained an export ban on their beloved cats for more than a thousand years. Indeed, there is little evidence of domesticated cats elsewhere in the world during this time. Domestic cats were mentioned in some Indian San-skrit writings from about 1000 B.C., but the Indians had almost certainly tamed their own breed of desert wildcat. And those domestic cats who do feature in Indian records of the period do not do so in anywhere near the profusion they do in Egypt.

When the Greeks began visiting and trading with Egypt, however, they soon recognized the superiority of cats to the ferrets and weasels they used to control mice and other rodents. Cats did not attack chickens and other valuable birds as frequently as weasels and they did not need to be caged when they were not catching vermin.

To get around the Egyptian export ban, the Greeks resorted to stealing several pairs of cats which they could then breed from themselves. The first record of a cat in Greece dates from about 500 B.C. A marble bas-relief shows a cat on a lead meeting a dog, also on a lead, and their owners who are apparently encouraging the two animals to fight. It is not clear, however, whether this cat was wild or had been bred in captivity.

Once the Greeks had produced a supply of cats sufficient for their own requirements, they began to offer cats to their trading partners across Europe. Thus the distribution of the domesticated cat began over 1,200 years after the Egyptians first accepted them into the human community. The expansion was slow and it took several hundred years before cats were bred in earnest.

Oddly, the Romans, who adopted so many practices and customs from the Greeks, do not appear to have incorporated cats into their daily lives until the fourth century A.D., as there are virtually no records or images of a domesticated cat before then. A first-century A.D. mosaic found in Naples shows a cat attacking a bird, but there is nothing to suggest that the cat had been tamed.

Even at Herculaneum and Pompeii, cities near Naples which were buried intact under lava in 79 A.D., not a single cat skeleton was ever found, nor did domestic cats feature in either town's numerous wall paintings and frescoes. One of the first records of domesticated cats among Romans comes from the fourth-century A.D. writer, Palladius. He recommended that cats be used for catching garden moles as a novel and hence uncommon alternative to polecats which were relatives of the ferret.

Once they recognized the value of cats as controllers of vermin, as well as household pets, the Romans often took cats with them throughout their empire. They may have been the first to introduce cats to Great Britain, for example. The oldest known evidence of a domesticated cat in Britain is a cat skeleton found among ruins of a Roman villa at Lullingstone in Kent which dates from the fifth century A.D.

By this time, however, man's perception of cats had changed dramatically from the Ancient Egyptian view of the cat as a demi-god. At the end of the fourth century A.D., Pope Theodosius I had officially outlawed Bastet and other pagan gods from the recognized realms of religion.

In its early years the Christian religion was fairly tolerant of cats and portrayed them favourably as the animal which was created by God to tackle the mouse, created by the Devil. According to one popular legend, when Noah prayed for the protection of his ark from the possible menace of mice, a lion sneezed and produced the first pair of cats from his nostrils. This was enough to send the mice scurrying into the nearest holes, their favourite place to hide ever since. Another legend, which originated in Italy, claimed that a cat was present in the stable at Bethlehem and gave birth to kittens at the same time that Mary gave birth to Christ. A cat was sometimes included in early paintings of the Holy Family and the Annunciation.

During the fifth century in Ireland a list of popular and 'useful items' for the Irish housewife was drafted. Cats were included in this list for the skill with which they caught mice as well as for their 'pleasant' disposition, making them far preferable to ferrets.

By the tenth century in west Wales, the cat's importance as a member of the community was confirmed by the official definition of a 'hamlet', which was described as ' . . . a place that contains nine buildings, one herdsman, one plough, one kiln, one churn, one bull, one cock and one cat'. The cat's value as a protector of the harvest was also officially recognized when Hywel the Good assigned a monetary value to cats in the year 936. A kitten was worth one penny before its eyes were open, which was the same price of a lamb, a hen or a goose. After its eyes were open, it was worth two pennies and it was worth four pennies once it could catch a mouse.

To protect cat owners, strict laws were enacted which forbade the killing of cats. Guilty parties had to pay the owner of a killed cat that cat's worth in corn. This was determined by hanging the dead cat by its tail until its nose just touched the ground, and then pouring corn over the cat's body until it was covered completely. By the twelfth century in Saxony, the fine for killing an adult cat had risen from this amount to sixty bushels of corn, the precious crop that cats protected so well.

During this time in the Far East cats were also prized by man, but for a different reason. When the Emperor of Japan decided to breed cats in the year 999 at the Imperial Palace in Kyoto, he began a fashion trend among the Japanese. Cats became prized curiosities and were walked on leads by proud owners. Many cats that were kept as pets were called 'tama', which meant 'jewel' in Japanese.

For several hundred years after the Emperor welcomed them to his palace, cats in Japan were kept primarily by the upper classes who bred them seriously and lavished attention on them. Cats were also kept indoors for the first time and few were left to attend to the mice at the silk factories or grain stores. For a while it was believed that an effigy of a cat was enough to keep the mice away, but this theory was soon proved wrong. By 1602 the government was forced to order its citizens to liberate their cats, imposing heavy fines on anyone who was caught buying or selling one, so that stores, factories and other industries could be protected against mice.

In Siam soldiers had found a different but equally important use for cats. Although it is hard to believe that cats would ever agree to do so, it is said that Siamese soldiers were able to train their cats to sit on top of their shoulders and spot enemies who tried to sneak up on them from behind. A cat would alert its soldier to an impending attack with special cries of warning.

Meanwhile, back in Europe, cats were fending off a growing number of attacks with little success, as they had fallen victim to some of the darker superstitious beliefs which flourished during the Middle Ages. The cat's more sinister and mysterious image, which had lain dormant for some time but was nevertheless a part of popular folklore, as well as of Christian legend, was being vigorously promoted by the Church.

In the thirteenth century, Europeans were becoming disillusioned with medieval society and with the Church which oversaw it. To fortify its waning power, the Church turned to witches as scapegoats for the ills in the world. A vicious and pervasive campaign was launched by the Church, claiming that witchcraft was the greatest threat to civilized society and all those who lived within it, rich and poor alike. Witches were subsequently hunted, purged and tortured by the Church for the next three hundred years across Europe and colonial North America.

In order to strengthen its conspiracy, the Church was able to draw upon the witch's long-standing and traditional association with the cat. In Roman mythology, cats had been connected with the goddess Diana, also known as Artemis, who was said to have created the cat to mock the lion which her twin brother, Apollo, had created. As the Roman poet Ovid wrote in his *Metamorphoses*, gods were able to turn themselves into animals whenever they had to flee from giants, and when the goddess Diana did so she turned herself into a cat. The cat was an appropriate symbol for Diana as she was known as the huntress-goddess.

Diana and, by turn, the cat, were also traditionally identified with the moon and moonlight. In this sense, Diana's image may have stemmed from the Ancient Egyptian image of Bastet, except that Diana's association with the cat and the moon eventually became quite sinister.

Diana was affiliated with the Roman goddess Hecate because she too was originally a moon-goddess. Hecate later became Goddess of the Underworld, however, but despite her change of status her image continued to overlap with that of Diana in popular mythology. To Diana and the cat were thus attributed many of Hecate's less desirable qualities and supernatural powers, such as the ability to enchant and possess people.

The association of Diana and her cat with the moon led them to be identified not only with Hecate and the Underworld, but also with the moon's dark phase and the evil deeds which occur on earth when it is cloaked in darkness. Both Diana and the cat were further associated with the moon's ability to change form.

The cat went here and there
And the moon spun round like a top,
And the nearest kin of the moon,
The creeping cat, looked up.

Black Minnaloushe stared at the moon,
For, wander and wail as he would,
The pure cold light in the sky
Troubled his animal blood . . .

Does Minnaloushe know that his pupils
Will pass from change to change,
And that from round to crescent,
From crescent to round they range?

Minnaloushe creeps through the grass
Alone, important and wise,
And lifts to the changing moon
His changing eyes.

W. B. Yeats (1865–1939)

Even people who were merely followers of Diana were
thought to be able to change form like the moon. Legends
subsequently developed about women who could change form
at night, usually becoming a cat, and such stories formed the
basis of later tales about witches.

Folk stories from around the world described women who
could transform themselves into cats at night, leave their beds
to roam and operate in their evil way, then change back into
women as they returned to bed at first light. Some of these
tales survived in parts of Europe for many years after the
Middle Ages.

A popular Spanish folk story from this century, for example,
tells of a man who put fresh milk outside his window at night
in order to keep it cool; in the morning he would often wake

to discover that the milk was gone. One night he waited patiently by the window to see who was stealing his milk. When a large black cat eventually appeared and began to drink the milk, he became furious and hit the cat's front leg with a stick. As it ran off into the night, the cat screamed with a voice that sounded very much like a human's. The next day, the man saw an old woman from a neighbouring village with one of her arms in a bandage. When asked what had happened, the woman replied she had fallen down some stairs; but the man knew right away that she was actually a witch who became a black cat at night and stole his milk.

Of course further 'evidence' of the existence of witches could be found in real life as well. As today, cats were often kept by elderly women who lived alone as they were relatively easy to look after and made good indoor companions. In most cases these women loved and protected their cats quite fiercely and would severely threaten anyone caught trying to harm or harass the cat. Naturally, when ill tidings *did* befall such a prankster, the 'wicked' old woman was frequently blamed. The prized cat, often seen stalking the streets at night, was assumed in many cases to be looking for revenge, either sent on the mission by the 'witch', or serving as a feline transformation of the wicked woman herself.

Thus, with an ample supply of material on hand with which to build its case, the Church was able to launch an extensive campaign against witches and their partners in crime, cats, portraying them as the Devil's allies and bearers of ill tidings to all men. In 1484 Pope Innocent VIII officially authorized the campaign when he publicly denounced cats and anyone who tried to protect them.

In a medieval book called *The Devil's Bible*, the cat's relationship with the Devil was described in some detail for the populace. All cats, for example, were said to be invited to dine with the Devil on Shrove Tuesday, which explained why it was unusual to see a cat on that day. Cats were said to cooperate

with the Devil in many ways, with some of their most important tasks occurring at night:

> Only imbeciles do not know that all cats have a pact with the
> Devil . . . You can understand why cats sleep or pretend to sleep
> all day long, beside the fire in winter or in the sun in the
> summer. It is their task to patrol the barns and stables all night,
> to see everything, to hear everything. And you can deduce from
> that why the Evil Spirits, warned [by cats] just in time, always
> manage to vanish away, to disappear before we can see them.

People who were in any way associated with cats, or could
be set up to look as if they were, were prime targets for the
Church's exhaustive witchhunts. In the 1600s, for example, a
Danish woman was accused of sorcery after giving birth to a
baby whose head looked like that of a cat. The baby most
likely suffered from a natural deformity of the brain called
anencephaly, but it was 'evidence' enough that the mother was
involved in witchcraft.

A woman who gave birth to twins was also suspect in some
countries where it was assumed that one of the pair would
have the ability to change into a cat. Children who kept cats or
were even seen playing with cats were often prosecuted. In
1699 over three hundred children were accused in the small
Swedish town of Mohra. Some were executed and others
tortured.

Indeed, torture induced many heretical 'confessions' from
people accused of witchcraft and these in turn fuelled the
public's belief in a witch epidemic. Cats were often included in
the torture and execution of witches; when a witch was burnt
at the stake, for example, cats would frequently be burnt with
her. And it was a widely held belief that as the fire consumed
the bodies of its victims, a black cat could be seen jumping out
from the flames.

Cats were also tortured independently of witches. They were
put to death in their hundreds after such natural disasters as a

flood, crop failure or fire, for which they were often blamed. Sometimes they were tortured or killed just for the sake of sport.

A popular activity during William Shakespeare's lifetime, for example, involved placing a cat in a bag or leather bottle and hanging it from a tree to be used as a target for archers. In *Much Ado about Nothing*, Shakespeare makes reference to this sport with the line 'Hang me in a bottle like a cat.' It was probably relatively easy for him to write this since it appears he was not a great cat lover. In all his work, he only made about forty references to cats, none of which were particularly favourable.

Cats were also put to death during the Middle Ages to mark certain religious occasions. Across Europe on St John's Day, which fell on Midsummer's Eve, cats were stuffed into baskets or sacks and then burned alive. The ashes were later taken home and kept as tokens of good luck. Although the Church had at one time tried to eliminate this traditional pagan ritual, which dated from the seventh century, by the Middle Ages it decided to incorporate it into its own doctrines. Because it had turned the cat into a symbol of the Devil, the Church taught that to burn cats in this way was actually acceptable since it purified communities of the Devil's evil power.

The public's fear and hatred of cats became so strong that it even infiltrated 'neutral' disciplines in society. In 1658, for instance, Edward Topsell completed a detailed scientific study in natural history which contained a lengthy chapter on the cat, its anatomy and behaviour. He concluded the chapter by writing 'The familiars of Witches do most ordinarily appear in the shape of cats, which is an argument that this beast is dangerous to soul and body.'

It was eventually believed that *all* aspects of a cat were harmful to mankind. Its skin was considered poisonous, as were its teeth, which were thought to be full of venom. Its breath could infect one's lungs and give one consumption. Even its hair was potentially fatal; a person who had the misfortune of swallowing just a small amount would certainly suffocate.

Thus the campaign against cats and witches continued to mount. In colonial America alone more than two thousand cat sorcery trials were recorded and for the first time in their history cats began to run the risk of extinction after evolving as a species for over fifty million years.

Finally, with the onslaught of the Black Plague in the 1660s, which killed half the inhabitants of London, cats were quickly reinstated into society to help wipe out the carrier of the disease, the black rat. It is perhaps ironic that the rat, which had always been hunted in such a skilful and merciless fashion by the cat, should have been responsible for the cat's salvation in Europe.

The French were among the first to recognize that cats, and the 'witches' they were presumed to represent, were *not* the cause of physical, mental and social ailments. Certainly they were among the first to declare the cat fashionable and to welcome it back into all levels of society.

In the early 1700s the Queen of France and noble ladies at the court of Louis XV pampered numerous cats. Animated discussions were held at fashionable salons about the cat's mysterious nature. Influential painters such as Fragonard, Boucher and Watteau began to include cats as part of their subject matter.

Elaborate tombs for beloved felines became a popular trend. Some cat lovers even made arrangements for their cat's care after their own deaths. Cardinal Richelieu, a Minister of State for France in the seventeenth century, kept dozens of cats during his lifetime and established a generous endowment for the fourteen who survived him.

In 1799 the French astronomer Joseph Jerome de Lalande decided that the cat deserved a place among the other great animals in the starry heavens, and he identified a new constellation which he called Felis. This might have been in answer to the eighteenth-century philosopher Voltaire, who had once asked, 'How can we be interested in an animal who did not know how to achieve a place in the night sky, where all the animals scintillate, from the bears and the dogs to the lion, the bull, the ram and the fish?' The cat's position in the heavens was only temporary though, for it was removed by astronomers less than one hundred years later when it was declared unjustifiable.

The cat enjoyed renewed popularity elsewhere in Europe during the eighteenth century. In England it was once again fully accepted into the home. This trend had apparently started a century or two beforehand, as confirmed by the Church reformer Desiderius Erasmus of Holland. After a visit in the sixteenth century, he complained in a letter to a friend that when visiting an English home one was required to kiss not only the host and hostess and all their children but the family's cat as well!

The English included the cat in many of their family activities and treated it as a well-respected member of the household. In a painting by William Hogarth of the Graham children in 1742, the family's tabby is included. 'Hogarth enjoyed painting the cat so much,' wrote the English art historian, Kenneth Clarke, '. . . that the Graham children look hollow and lifeless beside her. She is the embodiment of Cockney vitality, alert and adventurous – a sort of Nell Gwyn among cats.' Even King Charles I, who reigned from 1625 to 1649, kept a beloved cat which accompanied him almost everywhere. The cat was completely black and was considered by the King to be his token of good luck.

The association between cats and good luck stems, paradoxically, from the medieval belief that cats were possessed by the Devil. If a person came across a cat and was left unscathed, he or she was considered to have been unbelievably lucky. Similarly, if a cat came into one's home and was made to feel welcome and decided to stay, it was another sign of good luck;

by appeasing the cat, one appeased its master, the Devil, and it could not hurt to have the Devil on one's side.

An old saying in Britain may have also contributed to the general notion that cats can bring good luck, by claiming, 'Whenever the cat of the house is black, the lasses of lovers will have no lack.' Because a female cat on heat is able to attract a large number of interested toms, a house that kept such a cat was considered auspicious for single women interested in attracting male suitors.

In China cats were traditionally considered tokens of good luck. Shop owners often kept cats to bring their premises good fortune. The older and uglier the cat, the more luck it was supposed to bring.

In Europe black was considered luckiest for a cat because the colour was identified with the occult and associated supernatural powers. In America, however, the black cat was always considered demonic and it was the white cat, its exact opposite, which was thought to be lucky. 'The white cat,' writes Desmond Morris, 'presumably by direct contrast, was seen as a force of light against darkness and was in this way converted into a symbol of good fortune.' Even today Americans still consider the black cat to be unlucky whereas Europeans generally think the opposite.

I like black cats because:
They are discreetly dressed for every occasion
They bring good luck to their owners . . .
They are dramatic by day and invisible by night . . .

· Anonymous ·

The fate of King Charles I and his lucky black cat could serve as evidence that black cats can indeed bring good fortune to those who believe in their charms. After his black cat died, the King said, 'My luck is gone.' He probably had little idea how right he was for the very next day King Charles was arrested. He was held as a prisoner under orders of Oliver Cromwell and later beheaded.

Frederick the Great of Prussia, who died in 1786, also believed that cats were rewarding creatures to keep, but not necessarily because he considered them lucky. He appreciated their more practical attributes as reliable protectors of stored food. During the various military campaigns he undertook to expand his empire, Frederick the Great would order the citizens of every town he conquered to supply him with numerous cats. These were then officially designated guards of his army's food supply.

In other parts of the world the popularity of the cat had not waned to the degree it had in Europe during the Middle Ages, and by the end of the eighteenth century cats in these areas were still enjoying respectful and kind treatment.

Muslims, for instance, have always been fond of cats. Mohammed is said to have cut off the sleeve of his robe upon which his cat was sleeping rather than disturb it. His love of cats was well known throughout the Muslim world. Even today, as Mohammed's favourite animal, cats are still allowed inside mosques, where they are free to roam and never harassed.

In Asia cats were never victim of the conflict in Europe between the Christian Church and the Devil. As Adolph Suehsdorf wrote in a 1964 article for *National Geographic* magazine: 'Cats do not engage in witchcraft in areas where people do not believe in witches.'

By the nineteenth century in Victorian England, cats had become more popular than dogs as they were considered cleaner and safer to have as pets. Queen Victoria kept several cats during her lifetime, two of which were Blue Persians. When she died in 1901, she left her last cat, named White

Heather, to her son Edward VII who also kept several cats of his own.

The cat was still recognized for its practical value as a mouser of unmatched ability. In the latter half of the nineteenth century cats were given new opportunities to practise their skills as the Industrial Revolution had generated an exponential growth in the size of cities and the factories, warehouses and other facilities which supported them. Here mice and rats began to flourish on an unprecedented scale. To tackle the problem, managers of libraries, railway stations, warehouses and other affected public buildings recruited cats to live on the premises, fed on a stipend from the British government. The cats were very efficient and at one point over 100,000 were employed as British civil servants!

The largest corps of cats was probably maintained by the British Post Office which established the Cat System in 1868. The Post Office Secretary originally authorized the hiring of three cats to control mice, but agreed to pay only one shilling a week for the three cats' food instead of two shillings as suggested by the Money Order office. 'They must depend on the mice for the remainder of their emoluments,' he said, '. . . [and] if the mice be not reduced in number in six months a further portion of the allowance must be stopped.' When the system was seen to be a success, many other branches adopted the same policy. The average stipend of six or seven pence per cat was almost always too small to ensure that each one was fed properly. Thus, the number of mice was usually kept quite low. In 1873, though, a postmaster from Southampton was able to convince the government that he needed extra compensation for the support of the office cat, citing the wear on the leather of his shoes, as well as 'the loss of dignity, when carrying the cat's food through the streets in Her Majesty's uniform'.

But while cats were appreciated for reliable mousing skills and for their well-groomed appearance relative to other pets,

94

above all else they were especially popular in nineteenth-century England because they appealed to the Victorian's love of the beautiful and exotic.

Considered more attractive and graceful than dogs, cats were available in a variety of enticing foreign breeds. From the Orient a wide range had been exported, including the sleek Siamese cat from seventeenth-century Thailand, the plush Angora cat from sixteenth-century Turkey (named after the country's capital, Ankara, once pronounced *Angora*), and from the same period the extravagant Persian cat from Iran. Other breeds from the East included the Korat of Thailand, the Russian Blue, the Burmese and Birman of Burma, the Japanese Bobtail, the Abyssinian and the Egyptian Mau.

British cats, of course, were also kept but they were usually considered less exciting. Indeed, the West in general had fallen well behind the East in the development of pedigree cat lines. This was almost certainly due to the long dark period in the Middle Ages. Cats in the East, however, had been pampered and carefully bred for centuries. It is not surprising, therefore, that extravagant Eastern breeds became so popular in the West, especially with the aesthetically minded Victorian.

In the nineteenth century cat owners in Europe finally started to breed cats to enhance and refine their cat's particular features and characteristics. Cat shows were also organized for the first time, starting a tradition which continues to this day. The world's first cat show was held in Maine during the 1860s, followed by London's first cat show in 1871 at the Crystal Palace. As one might have suspected, Victorian values found their way into the classification of cats at these shows; both males and females were designated as such, but neuters were labelled 'Cats of No Sex', perhaps to pre-empt potentially embarrassing questions from children. One hundred and seventy cats participated in the first London show and were roughly divided into two groups, 'British' and 'Eastern'. These were then further categorized into twenty-five different classes.

The cats were ranked and judged by the show's organizer, Harrison Weir, his brother and the Reverend C. Macdona. An ex-cat hater, Mr Weir wrote an extensive book on the domesticated cat called *Our Cats and All about Them* in which he described cats as 'among animals the most perfect'. The book contained the first detailed descriptions and definitions of pedigree breeds and was described by the author as the 'outcome of over fifty years' careful, thoughtful, heedful observation, much research, and not unprofitable attention to the facts and fancies of others'.

This comprehensive book laid the foundation for societies of cat lovers which were soon formed in both America and Great Britain. Organizations were then established for the first time to protect cats, and other animals, from cruelty. The National Cat Club in Great Britain was founded in 1887 and operated its own Stud Book and Register of Cats. In 1910 this merged with its rival, the Cat Club, to form the Governing Council of the Cat Fancy, an organization which still exists today.

Now there are over one hundred recognized pedigree breeds of the domesticated cat. Across the world, the cat has become a well-established and popular housepet, outnumbering dogs in some countries. Many people would even go so far as to say a home is not complete without a cat.

A house without a cat, and a well-fed, well-petted,
and properly revered cat, may be a perfect house, perhaps,
but how can it prove its title?!

· Mark Twain (1835–1910) ·

Mark Twain was just one of many famous personalities who had well-known and happy relationships with cats. He was extremely fond of the animal and gave the cats in his household unusual names such as Zoroaster and Blatherstrike so that his children would be able to pronounce difficult words.

Jean Cocteau, the French poet and artist who lived from 1889 to 1963, also seemed to understand the contribution that cats can make to the atmosphere of one's home when they are correctly looked after. 'I love cats because I love my house,' he said, 'and little by little they become its visible soul. A kind of active silence emanates from these funny beasts . . . who move in a completely royal authority through the network of our acts, retaining only those that intrigue or comfort them.'

People known to have kept cats happily in their home have usually recognized and provided for their cat's various wishes. Sir Isaac Newton, for example, is credited with the invention of the cat flap, created specifically for the convenience of his cats.

Theodore Roosevelt, President of the United States from 1901 to 1909, had a six-toed grey cat named Slippers who loved to lie in the middle of a busy hallway between the East Room and a dining room at the White House. Such was Slippers' status in the House that anyone who came across him lying in the hall, including the President himself, was forced to walk around him.

The first U.S. President to keep a cat in the White House was Abraham Lincoln. He is known to have been extremely fond of cats and would go out of his way to be kind to them. On a visit to General Grant during the Civil War in the 1860s, for example, he spotted three kittens in a tent who had apparently lost their mother. Before leaving the campsite, he gave instructions that they were to be well looked after and fed, and he continued to inquire about their welfare on an almost daily basis. At the White House, Mr Lincoln enjoyed a rewarding relationship with his own pet cats which, he found, helped him to relax. According to his biographer, William Herndon, 'when . . . weariness set in he would stop thought, and get down and play with a little dog or kitten to recover'.

In Britain Sir Winston Churchill considered his cat to be an important member of the household. It is said that he let his cat sleep in the same bed and on occasion he put a chair at the dining table for it.

Thus, most people would probably agree that the relationship between domestic cats and their owners has now reached a comfortable and mutually rewarding level. Many might even say that cats now have the upper hand (or paw) around the house. They are as aloof and independent as they want to be, give out as much love and affection as they wish, and yet most still manage to have their owners cater to their every whim.

Could it be that man has adapted himself to the cat and not the other way round? Are we, like the Ancient Egyptians, happily subservient to the cat, expressing our adoration not through the construction of elaborate temples, but in ways which simply befit the times? Cats may indeed be as clever or cleverer than we think!

For not as a bond-servant or dependant has this proudest of mammals entered the human fraternity; not as a slave like the beasts of burden, or a humble campfollower like the dog. The cat is domestic only as far as suits its own ends; it will not be kennelled or harnessed nor suffer any dictation as to its goings out or comings in.
· Saki, *The Achievement of the Cat* ·

V
THE DEFINITIVE
CAT OWNER I.Q. TEST

Any conditioned cat-hater can be won over by any cat who chooses to make the effort.

· Paul Corey, *Do Cats Think?* ·

———————————

EACH CAT settles with its owner in a different way. Although generally independent, cats are also clever enough to understand that we can provide them not only with life's basic necessities but also some of its niceties. To get what they want out of life, cats must learn how to get what they want out of *us*.

Cat owners can be grouped roughly into two categories. There are those who recognize the difference between providing basic care for their cats and submitting to outright, though not always obvious, manipulation, and those who happily respond to their cat's every whim and enjoy doing so.

The Definitive Cat Owner I.Q. Test is designed to determine which of these two categories you fall into and, more specifically, the type of cat owner you are. The test contains seventy-five multiple choice questions which are grouped into four areas: Background, Training, Dedication and Sensitivity. Answer each question as accurately as possible, selecting only one response for every question.

If a question does not apply to you, try to determine the answer you would choose if the question *did* apply; if none of the possible answers apply, choose the one that best approximates to the answer you would like to give. Mark your selections as you go through the test for later reference to the scoring table.

After adding up all your points, refer to the Results Analysis section to convert your score to a Cat Owner I.Q. Once you have checked your I.Q. against the Cat Owner I.Q. Classification Table and discovered what type of owner you are, you can then consult the Recommended Breeds section to determine which breeds you might find most suitable.

HAVE FUN AND GOOD LUCK!

Part I Background

1. Did you have any cats when you were growing up and, if so, how many?

   A Yes, one to three. ☐

   B More than three. ☐

   C No, I did not have any cats. ☐

   D I knew a cat while growing up, but it was not mine. ☐

2. In general, how did you feel about cats when you were growing up?

   A I disliked them. ☐

   B Neutral. ☐

   C I liked them, but preferred other animals. ☐

   D I liked them better than most of the people I knew. ☐

3. Did you spend any significant time during your childhood with animals other than cats?

   A Yes. My family always had at least one dog. ☐

   B Yes. I had lots of different pets while growing up. ☐

   C Yes, but I never liked any other animal as much as cats. ☐

   D No. ☐

4. How would you compare cats with dogs?

A Some dogs, like some cats, are sweet and affectionate while others are cold and aloof; it depends on the breed and the animal's upbringing. ☐

B Dogs are much nicer, friendlier and more loyal than cats, and they make better pets. ☐

C Dogs may be friendlier and more loyal, but they are not nearly as mysterious, intelligent and intriguing as cats. ☐

The cat, an aristocrat, merits our esteem,
while the dog is only a scurvy type who got his
position by low flatteries.

· Alexandre Dumas (1812–70) ·

---

5. Do you believe that cats of a certain colour can bring a person bad luck?

A Yes. I think that black cats are unlucky. ☐

B Yes. I think that white cats are unlucky. ☐

C No. Cats do not affect a person's luck one way or the other. ☐

D No. I believe cats bring *good* luck to people. ☐

6. As a child, when others were unkind to a cat, how did it make you feel?

A Amused. ☐

B Saddened. ☐

C Very upset. I would do my best to stop it. ☐

D Neutral. I would have little or no reaction. ☐

7. When you see a cat at a friend's house or in the street, what do you usually do?

   A Approach it without hesitation and play with it for a while, if it's willing. ☐

   B Start speaking to it immediately in a high-pitched voice and try to pick it up and stroke it. ☐

   C Call for it to come to me and stroke it if it does, although in my experience cats hardly ever respond to strangers. ☐

   D Nothing, unless it comes over to me first, in which case I may stroke it. ☐

Those who'll play with cats must expect to be scratched.

· Cervantes (1547–1616) ·

---

8. Do you think cats make suitable pets for children?

   A Yes, if the cats have sweet dispositions. ☐

   B Yes. Cats won't put up with rough play and therefore they teach children how to handle animals. ☐

   C No. Cats have a greater tendency than most animals to scratch or bite when playing. ☐

9. How confident are you of your ability to befriend a cat?

   A Extremely confident as I've always been able to do so in the past. ☐

   B Not confident at all because it's always the cat who decides whether it wants to be friendly or not. ☐

   C Fairly confident, but some cats are just standoffish. ☐

To Someone very Good and Just
Who has proved worthy of her Trust
A Cat will sometimes condescend –
The Dog is Everybody's friend.

· Oliver Herford (1863–1935) ·

10. Do you believe, as some studies suggest, that keeping a cat as a pet is beneficial to one's health?

    A Yes, absolutely. ☐

    B In most cases, probably yes. ☐

    C No. In my experience they are generally too aloof to be good company. ☐

11. How do you feel about people when you discover that they do not like cats?

    A It makes no difference. ☐

    B I immediately lower my opinion of them. ☐

    C I feel sorry for them since they miss the joy of having a cat. ☐

    D I make a mental note to keep them away from my cat. ☐

Part IIa
How Well Have You Trained Your Cat?

12. When giving your cat a command, how often does it obey?

   A The situation wouldn't arise. I would never give my cat an order.

   B Hardly ever. 'Obey' is a four-letter word.

   C Almost always.

   D Sometimes, although I often sense it's because my cat wanted to do so anyway.

A dog knows its master, a cat does not.

· The Talmud ·

---

13. If your cat is sleeping on your bed and you move, your cat:

   A Acts annoyed and gives me a disgruntled miaow.

   B Spends the next five minutes rearranging its own sleeping position, making sure I know it has been disturbed.

   C Doesn't move in case it disturbs me.

   D May quickly rearrange its own position.

14. How often does your cat break the rules of household behaviour such as not using its litter box or not staying away from the rubbish bin?

   A Quite often. My cat has not fully accepted the idea of imposed behavioural limits.

   B Only occasionally and usually when it is upset about something.

   C Hardly ever, unless there is a very good reason for it.

   D Not applicable. I have not set any behavioural rules for my cat.

15. On average, when you come home and your cat greets you, which of the following positions is its tail in?

   A   (Very happy)

   B   (Pleased)

   C   (Nonchalant)

   D Tucked underneath its body as it is always sleeping.

16. Your cat is with you when you are eating something it particularly likes. If you left the room, with the plate on the table, your cat:

   A Would patiently wait at the table for me to return and offer it some food.

   B Would pace around the table and debate whether to wait for my return or to jump up and risk being reprimanded.

   C Would quickly jump on to the table, snaffle as much of the food as it could, then run away and hide.

   D Would leap on to the table as soon as I was out of sight, devouring the rest of the food on the spot.

17. If you were to leave a room in the identical circumstances described above and your cat did not give in to temptation, you would:

A Wonder if my cat was feeling all right. ☐

B Assume my cat was not hungry. ☐

C Reward its good behaviour with a share of the food. ☐

18. When you have friends to visit or hold a party, how does your cat behave itself?

A Like a spoiled child, jealous that it's not the centre of attention and annoyed at the disruption of its normal routine. ☐

B Perfectly well. It usually makes a cameo appearance, then disappears to another part of the house. ☐

C Sociably. My cat loves parties, mingling with the guests as if they've all come to visit *it*. ☐

D It takes advantage of the many opportunities to eat: enjoying offerings from the guests, morsels of food dropped on the floor or food placed on low-lying tables or unattended surfaces. ☐

19. You walk into the kitchen to find your cat on top of the work unit. Do you:

A Berate my cat for getting up there. ☐

B Gently tap my cat on the nose to let it know it's misbehaved, then put it down on the floor. ☐

C Simply stare at my cat and sternly call its name, causing it to scramble off the work surface. ☐

D Sweetly say hello and stroke it several times. ☐

20. How often does your cat respond when you call it?

    A Always. ☐

    B Hardly ever. ☐

    C Most of the time. ☐

    D Only if there is food involved. ☐

People with insufficient personalities are fond of cats.
They like being ignored.

· Henry Morgan (1915–   ) ·

Part IIb
How Well Has Your Cat Trained You?

Cats don't belong to nobody. General just rooms with me.

John Wayne, speaking of his cat, General,
in the movie *True Grit*

21. On average, how long does it take you to respond
to your cat's cry for food?

    A Until its next regularly scheduled feeding. ☐

    B Up to a few minutes, when I can get away from ☐
    my own activities.

    C A matter of seconds. ☐

22. How do you usually call your cat?

    A By whistling. ☐

    B By its name in a high-pitched voice. ☐

    C By its name with my voice at its normal pitch. ☐

    D By singing, 'Here, kitty, kitty!' ☐

23. What have you provided for your cat's instinctive
urge to scratch and sharpen its claws?

    A Nothing because my cat only rarely scratches ☐
    anything.

    B I have my cat's claws clipped frequently. ☐

    C I bought it a scratching-post. ☐

    D I let it scratch whatever it wants, including ☐
    furniture and curtains.

24. How do you feel about giving your cat food from the table?

A I occasionally give in if asked, even though I tell myself it's the last time. □

B I always give in if asked, knowing very well it won't be the last time. □

C I give the cat food even if it *doesn't* ask for it. □

D I never give it food from the table because I believe it is spoiling it to do so. □

The kind man feeds his cat before sitting down to dinner.

· Hebrew proverb ·

---

25. How often do you find yourself sharing your thoughts with your cat?

A Not very often. □

B Only if I need to discipline it verbally. □

C Almost every day. □

D Many times a day. □

26. When you see your cat chewing a plant or grass, what do you think this indicates?

A It did not get enough food at its last meal. □

B It is not feeling well and craves certain nutrients in the plant. □

C It is becoming a vegetarian. □

D It has an appetite which never ceases to amaze me. □

27. How often does your cat seem to get its own way?

  A Almost all the time. □

  B All the time. □

  C Most of the time. I occasionally deny it what it
    may want, such as yet more food or my side of
    the bed. □

  D Only some of the time. I am quite disciplined
    with my cat. □

Cats know how to obtain food without labour,
shelter without confinement, and love without penalties.

· Walter Lionel George (1882–1926) ·

---

28. Which of the following cat sounds do you think you
  can make quite convincingly?

  A A miaow. □

  B A hiss and a miaow. □

  C A hiss, miaow and growl. □

  D A hiss, miaow, growl and a scream. □

29. How many special gifts have you bought for your
  cat?

  A Countless. □

  B More than I've bought for myself or any other
    person I know. □

  C Several. □

  D None. □

30. How well do you plan ahead to ensure that there is always enough food for your cat?

A I am extremely well-organized and always have at least a week's supply on hand. ☐

B I rarely forget to buy it food, but, if I do, I simply give it some of mine. ☐

C I forget to buy its food fairly often, so my cat gets to share my food frequently. ☐

D I don't need to plan ahead as my cat always eats whatever I have. ☐

31. What do you do when your cat gets sick?

A Try to diagnose the problem myself, calling the vet if necessary. ☐

B Call the vet immediately and follow his or her instructions to the letter. ☐

C Wait for my cat to get better, only calling the vet if the condition is very serious. ☐

32. When you speak to your cat, do you address it as if:
A It is a newborn baby. ☐
B It is a six-year-old child. ☐
C It is your own age. ☐
D It is older and wiser than you. ☐

33. Your cat is sleeping peacefully and you need to wake it up for some reason. Do you:

A Yell at it from across the room. ☐

B Stroke it and whisper its name, apologizing for the inconvenience. ☐

C Gently shake it and ask it to wake up. ☐

34. How affectionate are you with your cat?

   A Reasonably affectionate. ☐

   B Not very affectionate. ☐

   C Pretty affectionate. I like to hold and stroke my cat as often as I can. ☐

   D I would like to be more affectionate, but my cat is aloof. ☐

35. If you've given your cat its usual amount of food, and it miaows for more, do you:

   A Give it as much as it wants. ☐

   B Give it *some*, for fear of spoiling it. ☐

   C Make a mental note to advance the time of its next feeding. ☐

   D Ignore the request altogether because I know it's had enough food. ☐

36. How often do you groom your cat?

   A Daily. ☐

   B Once a week or so. ☐

   C Rarely. It doesn't seem to require grooming very often. ☐

   D Never. It does its own grooming. ☐

37. If your cat suddenly jumps on to your bed while you're trying to sleep, do you:

A Let it go wherever it pleases, even if it wants the spot I occupy. ☐

B Pay little attention because my cat knows it must be considerate. ☐

C Watch while my cat walks over my chest or head, preparing to settle there. ☐

Part III Dedication

---

38. How often do you go out of your way to do something nice for your cat?

  A Hardly ever. ☐

  B A couple of times a year. ☐

  C Once or twice a month. ☐

  D Almost every day. ☐

I never shall forget the indulgence with which he treated Hodge, his cat; for whom he himself used to go out and buy oysters, lest the servants having that trouble should take a dislike to the poor creature.

· Mr Boswell, speaking of Dr Samuel Johnson (1709–84) ·

---

39. How many photographs do you have of your cat?

  A One or two. ☐

  B Several, with some framed and placed among photos of friends and family. ☐

  C Several, but none are displayed in the house. ☐

  D None at all. ☐

40. When you're playing with your cat, how long can you continue the game?

  A Until the cat gets bored. ☐

  B Not applicable. I don't like to play games with cats. ☐

  C A minute or two. ☐

  D Up to ten minutes. ☐

When I play with my cat, who knows whether
she is not amusing herself more with me than I with her.
We entertain each other with mutual follies.

· Michel de Montaigne (1533–92) ·

41. For some reason, your cat decides not to use its litter box but a sofa or chair instead. How would you feel about this?

A Annoyed, but I would ask the cat not to do so again in a very gentle manner.

B Quite annoyed, being firm but kind while reprimanding it.

C Irritated, but aware that the cat might be trying to tell me something.

D Furious, akin to the sentiments of Mrs Thomas Carlyle as quoted below:

> ... that cat!! – I wish she were dead! But I can't shorten her days! because ... as long as she attends Mr C. at his meals (she doesn't care a snuff of tobacco for him at any other times!) so long will Mr C. continue to give her bits of meat, and dribbles of milk, to the ruination of carpets and hearth-rugs!!
>
> Mrs Thomas Carlyle, wife of the author,
> from a letter to her maid, 1865

42. When you call your cat to come to you, or jump into your lap, and it won't, how long do you continue calling?

   A Longer than I care to admit. ☐
   B About a minute. ☐
   C A good five minutes or so, if I have it to spare. ☐
   D I stop calling after two or three attempts. ☐

43. Would you ever seriously consider taking your cat to a cat psychologist?

   A Probably not. ☐
   B Absolutely not. I would never consider it. ☐
   C Possibly, but only if I had a serious problem with my cat. ☐
   D Yes. I have done so in the past or plan to in the future. ☐

44. If you were to enter a room and saw an adorable kitten, would you:

   A Immediately start to stroke the kitten, probably forgetting why I had come into the room. ☐
   B Finish whatever it was I had come to do and then stroke the kitten. ☐
   C Stroke the kitten for a little while first, but then carry on with my own activities. ☐
   D Ignore the kitten altogether. ☐

. . . the playful kitten, with its pretty little tigerish gambols, is infinitely more amusing than half the people one is obliged to live with in the world.

· Lady Morgan, Irish authoress (1783–1859) ·

---

45. How do you react if your cat starts miaowing for no apparent reason, waking you in the middle of the night?

A Very irritated.

B Concerned.

C Slightly annoyed.

D Realize that it wants to snuggle up with me in bed.

46. In 1553 the *Fila Caverna* of Venice sailed to Jerusalem and a passenger on the boat wrote the following account of an incident involving the Master of the Ship's cat:

> . . . It chanced by fortune that the shippes Cat lept into the sea, which being downe, kept her selfe very valiauntly above water, notwithstanding the great waves, still swimming, the which the master knowing, he caused the Skiffe with halfe a dozen men to goe towards her and fetch her againe, when she was almost halfe a mile from the shippe, and all this while the shippe lay on staies. I hardly believe they would have made such haste and meanes if one of the company had bene in the like perill.

If you were in the same situation, would you:

A Do the same thing as the Master of the Ship.

B Send not six, but twenty men to save my cat.

C Try to steer the ship nearer the cat, then send one man to save it.

D Throw the cat a raft with a food supply, if it couldn't be reached.

47. When passing a pet shop with time on your hands, do you:

A Only go in if the shop looks interesting. ☐

B Usually walk past. ☐

C Almost always go in and buy something for my cat. ☐

D Usually go in, occasionally getting an item for my cat. ☐

48. On an average day, how long is it before you greet your cat when you get home?

A No time at all. The first thing I do is look for my cat. ☐

B Five minutes or so, after I've put my coat and keys away. ☐

C Quite a while because my cat likes to hide. ☐

D It varies tremendously since my cat's whereabouts are so unpredictable. ☐

49. How well do you feed your cat?

A Very well, sometimes sharing my food with it. ☐

B Quite well; I don't often share my food with it, but it can have as much of its own food as it wants. ☐

C Better than I feed myself. My cat has its favourite food *daily*. ☐

D Very well and very carefully. My cat has a regular feeding schedule and a specifically designed, balanced diet. ☐

Those who feed cats well will have sun on their wedding day.

· Welsh superstition ·

---

50. How many cat-related household items, such as pictures, ceramics and decorative accessories, do you have in your home?

A None.

B One or two.

C Up to ten.

D Many more than ten. My home is a tribute to felines everywhere, especially my own.

51. Do you buy your cat something special for its birthday, Christmas or other special occasions?

A Yes. My cat is treated like any other member of the family in this respect.

B No. Cats cannot appreciate the significance of such gifts.

C Although I may remember my cat's birthday, I don't buy it anything to celebrate it.

D I'm not sure when my cat's birthday is, but I always include it in my own special celebrations, giving it extra food or buying it a special treat.

52. What arrangements do you make for your cat when you go away?

A I simply leave out enough food and drink to last for the duration.

B I arrange for someone to stop by daily to feed and look after it.

C I book it into a plush cattery.

D I take it with me wherever I go.

53. When inside a pet shop, do you:

   A Avoid the cats for sale section as I can't stand to see them cooped up. ☐

   B Rush immediately to the cats section, wishing I could buy every one in sight. ☐

   C Occasionally look at the cats for sale and choose one that I'd like to have myself. ☐

   D Purchase what I came into the shop to buy, then leave immediately. ☐

54. How often do you take your cat to the vet, excluding emergency and other unplanned visits?

   A Once every three months. ☐

   B Twice a year. ☐

   C Once a year or less often. ☐

55. Do you ever do something special for your cat after you've taken it to the vet?

   A No, because visits to the vet are expensive and I feel it's quite enough taking it there regularly. ☐

   B Sometimes, if the visit has been particularly harrowing. ☐

   C Rarely because I don't want to spoil my cat. ☐

   D Yes, I feel guilty about taking it there and treat it to its favourite food. ☐

56. How often do you initiate playful games with your cat?

   A Several times a day. ☐

   B At least once a day. ☐

   C Hardly ever. ☐

   D I've given up trying as my cat rarely wants to play. ☐

57. How often do you talk about your cat to other people?

A Quite often, if they are interested in hearing my stories. ☐

B Quite often, even if they're *not* interested in hearing my stories. ☐

C I only talk about my cat if someone else asks me about it. ☐

D Hardly ever and only if there is something particularly important or funny to say about my cat. ☐

58. How do you like listening to other people tell stories about their cat?

A I usually find it interesting and amusing, if they don't go on and on. ☐

B I like to listen to others' cat stories very much and compare them to my own. ☐

C I don't mind listening at all, but I never find their stories as interesting as my own. ☐

D I find it incredibly tedious and boring. ☐

Part IV Sensitivity

59. If you were to accidentally step on your cat, hurting it, how would you react?

   A With annoyance that my cat was once again underfoot. ☐

   B Guilty for having stepped on it, apologizing right away. ☐

   C Terribly guilty. I would try to make up for it by stroking it, if it would let me, or, more appropriately, offering some favourite food. ☐

   D I would make sure that the cat was not seriously injured, then carry on with my own activities. ☐

60. Your cat dozes off near the television when you decide you would like to watch a programme. Would you:

   A Turn on the television, but keep the volume low. ☐

   B Not turn on the television. ☐

   C Watch my programme at the usual volume. ☐

   D Gently move my cat to where it can sleep peacefully. ☐

61. If you were to say something negative about your cat, and you think that it may be in earshot, do you:

A Feel it doesn't apply. I would never say anything negative about my cat in the first place. ☐

B Switch to positive, loving statements about it. ☐

C Continue my conversation. ☐

D Change the subject abruptly. ☐

I recollect Hodge (the cat) one day scrambling up Dr Johnson's breast, apparently with much satisfaction, while my friend, smiling and half-whistling, rubbed down his back, and patted him by the tail; and when I observed he was a fine cat, saying, 'Why, yes, Sir, but I have had cats whom I liked better than this'; and then, as if perceiving Hodge to be out of countenance, adding, 'but he is a very fine cat, a very fine cat indeed'.

Mr Boswell,
speaking of Dr Samuel Johnson (1709–84)

---

62. Cats make different sounds for different occasions. How many of the following types of miaows are you able to identify?

(1) Greeting miaow.　　(2) Hungry miaow.
(3) Angry miaow.　　(4) Pleased miaow.

A One. ☐

B Two. ☐

C Three. ☐

D Four. ☐

63. Do you worry about disciplining your cat in front of others?

  A Yes. I'd never do so in front of others, out of respect for the cat. ☐

  B I would only think twice if I needed to discipline it severely. ☐

  C No, I don't worry about it because a cat should be disciplined for a wrongful deed as soon as it occurs, otherwise it won't understand what it has done wrong. ☐

  D I never discipline my cat. ☐

64. What do you do if your cat tries to bite you?

  A Smack it lightly and tell it never to do that again. ☐

  B Scold it verbally. ☐

  C Give it a toy to bite on instead. ☐

  D Recognize that it's trying to tell me something and attempt to find out what that is. ☐

According to cat therapist Carole Wilbourn of New York, cats who bite are often suffering from Single Cat Syndrome. '[Biting] is an attention-getting behaviour caused by boredom,' she says.

---

65. When you are busy at home and hear your cat miaowing in another room, do you:

  A Try to interpret the nature of the miaow before deciding what to do about it. ☐

  B Drop everything and run to see if it's all right. ☐

  C Continue with what I was doing unless it sounds like an emergency. ☐

66. Your cat is particularly chatty and you suspect that it is trying to tell you something. Do you:

    A Ignore it. ☐

    B Try to understand, making responsive sounds. ☐

    C Take it in my arms and/or follow it until I discover what it is. ☐

    D Tell it to go away and be quiet. ☐

67. What kind of sleeping arrangements have you provided for your cat?

    A I bought the most luxurious cat bed I could find and/or afford. ☐

    B I bought an average cat bed that is functional but not distinctive. ☐

    C None, because my cat already had a bed – mine. ☐

    D No provisions are necessary because my cat can sleep anywhere, at any time. ☐

68. Do you enjoy having the cat on your lap or by your side when you are sitting down or sleeping?

    A Yes, very much. ☐

    B Yes, most of the time. ☐

    C Sometimes. ☐

    D No, I prefer it to keep its distance. ☐

69. What do you think it means when your cat's ears are pulled down flat over its head?

    A It is cold. ☐

    B It is fearful or angry about something. ☐

    C It wants some peace and quiet. ☐

    D It is exercising its ears. ☐

70. Cats often bring their owners presents, such as captured birds or a recently chewed piece of their clothing. If your cat were to bring you such a present, how would you react?

A Happy that my cat wants to please me. ☐

B I would acknowledge my cat's gift, but reprimand it gently for the damage it had done. ☐

C Annoyed and bothered by its action, scolding it right away. ☐

71. If you overhear a friend verbally disciplining your cat, how does it make you feel?

A Concerned for my friend. He or she would only be acting fairly and the situation must have called for it. ☐

B Angry at the cat for misbehaving towards a friend. ☐

C Annoyed at my friend for stepping in. ☐

D Quite angry that my friend had stepped in. ☐

I once chid my wife for beating the cat before the maid, who will now, said I, treat puss with cruelty perhaps, and plead her Mistress's example.

· Dr Samuel Johnson, critic and writer (1709–84) ·

---

72. When you're about to sit down to a meal and your cat is watching, do you:

A Prepare a small bowl of my food for the cat. ☐

B Prepare a small bowl of cat food, even if it is not its mealtime. ☐

C Prepare a bowl of food for my cat only if it's mealtime. ☐

73. Your cat cries persistently and is obviously upset about something. You become:

   A Convinced that cats are completely unsuitable as housepets.

   B Upset until your cat feels better and stops crying.

   C Concerned.

   D Irritated.

74. When you are trying to work or study at home and your cat will not leave you alone, do you:

   A Patiently wait for my cat to calm down, while trying to continue my work.

   B Abandon my project temporarily and pay attention to the cat, until it seems satisfied and leaves me alone.

   C Gather up my work and move to another part of the house, hoping my cat won't follow.

   D Lock the cat out of the room.

75. Your cat is sleeping in a chair you need to occupy. Do you:

   A Attempt to move my cat to another chair without waking it up.

   B Toss my cat off the chair.

   C Softly call my cat's name and wake it up gently.

   D Re-assess my need to occupy the chair.

SCORING TABLE

Part I Background

Question 1	**Question 2**	**Question 3**	**Question 4**
A = 3	A = 1	A = 2	A = 2
B = 4	B = 2	B = 3	B = 1
C = 1	C = 3	C = 4	C = 3
D = 2	D = 4	D = 1	

Question 5	**Question 6**	**Question 7**	**Question 8**
A = 1	A = 1	A = 3	A = 2
B = 1	B = 3	B = 4	B = 3
C = 2	C = 4	C = 2	C = 1
D = 3	D = 2	D = 1	

Question 9	**Question 10**	**Question 11**	
A = 3	A = 3	A = 1	
B = 1	B = 2	B = 2	
C = 2	C = 1	C = 3	
		D = 4	

Part IIa
How Well Have You Trained Your Cat?

Question 12	**Question 13**	**Question 14**	**Question 15**
A = 4	A = 3	A = 3	A = 4
B = 3	B = 4	B = 2	B = 3
C = 1	C = 1	C = 1	C = 2
D = 2	D = 2	D = 4	D = 1

Question 16	Question 17	Question 18	Question 19
A = 1	A = 3	A = 4	A = 1
B = 2	B = 1	B = 1	B = 3
C = 3	C = 2	C = 2	C = 2
D = 4		D = 3	D = 4

Question 20

A = 1
B = 3
C = 2
D = 4

Part IIb
How Well Has Your Cat Trained You?

Question 21	Question 22	Question 23	Question 24
A = 1	A = 2	A = 2	A = 2
B = 2	B = 4	B = 1	B = 3
C = 3	C = 3	C = 3	C = 4
	D = 1	D = 4	D = 1

Question 25	Question 26	Question 27	Question 28
A = 2	A = 2	A = 3	A = 1
B = 1	B = 4	B = 4	B = 2
C = 3	C = 1	C = 2	C = 3
D = 4	D = 3	D = 1	D = 4

Question 29	Question 30	Question 31	Question 32
A = 4	A = 1	A = 2	A = 1
B = 3	B = 2	B = 3	B = 2
C = 2	C = 3	C = 1	C = 3
D = 1	D = 4		D = 4

Question 33	Question 34	Question 35	Question 36
A = 1	A = 3	A = 4	A = 4
B = 3	B = 1	B = 3	B = 3
C = 2	C = 4	C = 2	C = 2
	D = 2	D = 1	D = 1

Question 37

A = 2
B = 1
C = 3

Part III
Dedication

Question 38	Question 39	Question 40	Question 41
A = 1	A = 2	A = 4	A = 4
B = 2	B = 4	B = 1	B = 3
C = 3	C = 3	C = 2	C = 2
D = 4	D = 1	D = 3	D = 1

Question 42	Question 43	Question 44	Question 45
A = 4	A = 2	A = 4	A = 1
B = 2	B = 1	B = 2	B = 3
C = 3	C = 3	C = 3	C = 2
D = 1	D = 4	D = 1	D = 4

Question 46	Question 47	Question 48	Question 49
A = 3	A = 2	A = 4	A = 3
B = 4	B = 1	B = 3	B = 2
C = 2	C = 4	C = 2	C = 4
D = 1	D = 3	D = 1	D = 1

Question 50	Question 51	Question 52	Question 53
A = 1	A = 4	A = 1	A = 3
B = 2	B = 1	B = 2	B = 4
C = 3	C = 2	C = 3	C = 2
D = 4	D = 3	D = 4	D = 1

Question 54	Question 55	Question 56	Question 57
A = 3	A = 1	A = 4	A = 3
B = 2	B = 3	B = 3	B = 4
C = 1	C = 2	C = 1	C = 1
	D = 4	D = 2	D = 2

Question 58

A = 2
B = 4
C = 3
D = 1

Part IV Sensitivity

Question 59	Question 60	Question 61	Question 62
A = 1	A = 2	A = 4	A = 1
B = 3	B = 4	B = 3	B = 2
C = 4	C = 1	C = 1	C = 3
D = 2	D = 3	D = 2	D = 4

Question 63	Question 64	Question 65	Question 66
A = 3	A = 1	A = 2	A = 1
B = 2	B = 2	B = 3	B = 3
C = 1	C = 3	C = 1	C = 4
D = 4	D = 4		D = 2

Question 67

A = 3
B = 2
C = 4
D = 1

Question 68

A = 4
B = 3
C = 2
D = 1

Question 69

A = 2
B = 4
C = 3
D = 1

Question 70

A = 3
B = 2
C = 1

Question 71

A = 1
B = 2
C = 3
D = 4

Question 72

A = 3
B = 2
C = 1

Question 73

A = 1
B = 4
C = 3
D = 2

Question 74

A = 3
B = 4
C = 2
D = 1

Question 75

A = 3
B = 1
C = 2
D = 4

RESULTS ANALYSIS

Cat Owner I.Q. Conversion Graph

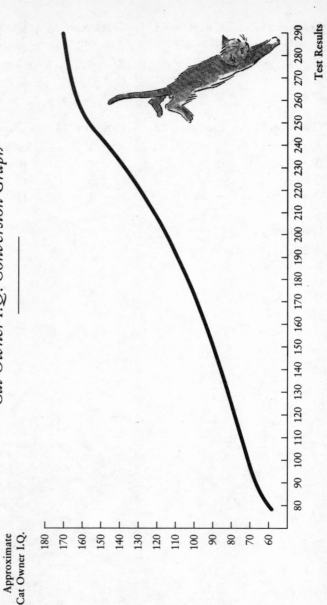

Approximate
Cat Owner I.Q.

Test Results

Cat Owner I.Q. Classification Table

CAT OWNER I.Q.	OWNER TYPE
80 and below	Practical
81–100	Flexible
101–139	Congenial
140 and above	Fanatical

RECOMMENDED BREEDS BY OWNER TYPE

Type 1: Practical (I.Q. ≤ 80)

Summary   As a Practical owner, you need a cat that is highly intelligent and highly independent. Your ideal cat will take care of itself and does not want an excessive amount of attention lavished upon it. Indeed, it considers most displays of affection a nuisance.

Care   Your cat should be adept at handling its own grooming and, ideally, its own feeding schedule. You would like to leave all its food out for the entire day without worrying that your cat might eat too much or too little.

Features   Short hair is a must as you would not want to be bothered with the necessary brushing of a long-haired cat, nor with long cat hairs that would be left everywhere. An exotic breed might appeal to you, especially one that adorns your home with its good looks.

Environment   You need an indoor cat, one that is perfectly content to stay inside. You do not want to feel pressured to let your cat in and out of the house or, worse still, to spend inordinate amounts of time searching for it up and down the street.

Your ideal cat should be able to find its own means of amusement around the house, preferring these to any special cat toys which you probably would be reluctant to buy anyway!

Recommended Breeds:

BOMBAY

BRITISH SHORT HAIR

EUROPEAN AND AMERICAN SHORT HAIR

REX

RUSSIAN BLUE

SCOTTISH FOLD

SINGAPURA

Type 2: Flexible (I.Q. 81–100)

Summary Flexible owners often think of themselves as Practical when, in actual fact, they give in to their cat's wishes more often than they realize. This usually happens unknowingly, and almost always at the instigation of the cat concerned. Flexible owners are thus able to care very well for the same cat breeds as Practical owners, but they are better suited to cats which appreciate and ask for affection now and then.

Care You would be happiest with a cat that does most of its own grooming, but requires some brushing and the occasional bath. Chances are that you will enjoy doing these things for your cat when prompted. Your ideal cat is happy with a balanced and regular diet, but will sometimes request a little variety. While clever enough to pace itself throughout the day, if necessary, it would prefer you to personally set out its food at each feeding.

Features Short-haired breeds are preferable to long-haired cats as they are less trouble to keep groomed. You may find that you feel closer to a cat which you consider visually attractive. Likewise, a cat that is inherently sweet, communicative and often affectionate will make sure that you remain involved with it.

Environment An indoor cat, one that rarely wishes to go outside, will be happy with you. Your ideal cat would enjoy having a wide variety of objects in the house to play with as it will want to initiate games with you. As your cat likes to communicate, a household with several family members or frequent social gatherings would be nice.

Recommended Breeds:
BIRMAN
BRITISH BLUE
CORNWALL MANX
EGYPTIAN MAU
KORAT
OCICAT
SHORT HAIR PERSIANS

Type 3: Congenial (I.Q. 101–139)

Summary   As a Congenial cat owner, you are happy to do almost anything your cat wants or needs. However, unlike a Fanatical cat owner, you rarely consider what else you could do for your cat that it hasn't already asked for. In fact, you may limit unsolicited acts of kindness for fear of spoiling your cat.

The more time your cat spends with you the better, as you consider it a great companion. Happiest when you feel that you have a strong relationship with your cat, you are best suited to a very communicative cat which needs a lot of attention, but is not a burden.

Care   You will probably enjoy a cat that needs grooming on a regular basis. Your cat should let you know whenever it is ready to eat and will ideally join you at most of your meals. While not too fussy about what it eats, your ideal cat will enjoy a wide range of foods.

Features You would be happy with either a short-haired or long-haired breed, but cats with long hair are preferable as they need to be groomed more often. A cat that is affectionate, as well as playful, and genuinely likes to be with you and others will be the most rewarding.

Environment Your ideal cat will be happy to spend most of its time indoors, but it also enjoys going out every now and then, especially if you go too. Always able to keep itself amused, it would nevertheless appreciate any item you buy for it. Your ideal cat likes social gatherings and would prefer a household in which there are several members.

Recommended Breeds:

BALINESE

BURMESE

HAVANA BROWN

LONG HAIR PERSIANS

MAINE COON

RAGDOLL

SPHYNX

TURKISH VAN CAT

Type 4: Fanatical (I.Q. ≥ 140)

Summary   As a Fanatical cat owner, the world revolves around your cat. It is given free rein over everything in the house, including people. Nothing is too much trouble and, indeed, to please your cat pleases you. Although your ideal cat may be quite capable of taking care of itself, you will be convinced that it can't live without you.

Care   Your cat must enjoy being groomed as you will probably brush it every day and bath it nearly as often, whether it needs it or not. Although its food dish is almost certainly never empty, your ideal cat will expect to be served each of its meals and will want to be included in yours as well. It will prefer a rich and varied diet.

Features A high-maintenance cat, one that wants to be fussed over all the time, is the most compatible with Fanatical owners. Your ideal cat will have many requirements and create countless opportunities for you to try to please it.

Long-haired breeds are certainly preferable to short-haired breeds as they require more brushing, bathing and grooming. Likewise, a cat that is less intelligent than average or one that is incredibly finicky will need your attention more often. For the Fanatical owner, looks are often less important than personality; the most enjoyable cat will be very demanding but demonstrably appreciative of your attention and care.

Environment Willing to do almost anything to make your cat happy, you want to create the perfect environment for its relentless pursuit of absolute self-gratification. Your ideal cat will appreciate any gifts you buy for it, but will become bored with them quite quickly, prompting you to find new ones, something you like to do.

Everyone in your home should be made aware of your cat's special status as household hedonist. There should be a quiet spot somewhere in the house, as well as food and drink in its bowls at all times. Both indoor cats and outdoor cats are equally suited to you, but your ideal cat will want to enjoy both environments, usually on a daily basis.

Recommended Breeds:

ABYSSINIAN

ANGORA

HIMALAYAN

JAPANESE BOBTAIL

MALAYAN

SIAMESE

SOMALI

TONKINESE

VI
INTELLIGENCE
ISN'T EVERYTHING

———

———

A FTER YOU have determined the I.Q.s of your cat and of yourself as cat owner, it is important to remember that many different factors contribute to the success of a relationship between a cat and its owner, most of which are more important than respective levels of I.Q. While it is amusing to speculate on the intelligence of your favourite feline, as well as on your own level of understanding and ability to care for it as a cat owner, assessing the I.Q. of yourself and your cat is meant to be an entertaining and insightful exercise rather than a clinical and scientifically accurate one.

Respective levels of I.Q. need not correspond to ensure a happy relationship. A very intelligent cat might be compatible with someone who has a high, 'fanatical' cat-owner I.Q., although it is more likely to be suited to someone with a low, 'practical' I.Q. score. The same is true for cats that did not score highly on the I.Q. test. While they are probably most compatible with high-I.Q., Fanatical cat owners who dote on them excessively, they might also get along with, or even prefer, a cat owner who earned a lower, more 'practical' score.

Thus, in defining the variables that make the relationship between a cat and its owner successful, one must look beyond levels of intelligence and I.Q. and consider other, more significant, factors.

UPBRINGING

One of the most critical factors is the successful socialization of the cat, teaching it how to interact with those around it, which is especially effective at a young age. Just as kittens who are handled frequently and given extra care during the first few weeks of their lives are more likely to be intelligent, those introduced to the sounds, activities and people in their environment at a young age are more likely to adapt to and feel comfortable with their domestic lifestyle.

The most critical period in a cat's life for learning how to relate to its environment and those who share it seems to be between the ages of four to eight weeks. Introducing a kitten to standard household procedures such as the use of a litter box, the noises of domestic appliances, and brushing and other methods of grooming, will encourage it to accept such activities as it grows.

If it is regularly held and gently played with as a kitten, it is also more likely to enjoy affection and handling as an adult. A well-adjusted cat who is comfortable in its environment and at ease with other people has a greater chance of developing a rewarding relationship with its owner.

COMMUNICATION

Another important factor in the success of the relationship is the ability to communicate effectively. For this to happen, each side should understand a majority of the signals that the other employs to convey thoughts. The eyes say a lot about a cat's mood, for example. The pupils change size not only in reaction to changes in light or proximity to the object in focus, but also in reaction to any strong emotions the cat might experience. Watch your cat's eyes the next time you give it food. Upon seeing the food, assuming it is hungry, the pupils of your cat's eyes will probably expand dramatically; in some cats, they become four to five times larger than they were in the space of just one second. And the pupils will also enlarge when the cat sees something threatening. A quick change in the size of the pupils could indicate either a strong positive or negative emotion, and one must therefore interpret the signal in context.

The degree to which the eyes are open is another indicator of a cat's current state of mind. If the eyes are open wide, the cat is alert or perhaps on guard. This is often the case when it hears unfamiliar noises or when it is surrounded by strangers or people it does not trust. If the eyes are only half-way open, though, the cat is usually satisfied and completely relaxed.

Staring is one of the strongest signals a cat can make with its eyes. Developed as an integral part of its predatory behaviour, hard staring is an instinctive and important method of intimidating an opponent. Cats almost always mean it as a sign of aggression and if they receive a stare, it is usually interpreted as a threat.

Perhaps the most obvious and direct way a cat can express itself, though, is by drawing upon its wide range of vocal sounds. Unlike dogs who use only vowel sounds when they speak, cats are said to use both vowel and consonant sounds to express their thoughts. Researchers agree that the range of feline vocal sounds is extensive, with some even claiming it as the most elaborate sound range of any animal other than man.

Even within a single type of sound, the variety of tones and messages that a cat can convey is quite impressive. The nineteenth-century writer Champfleury, for whom cats were a favourite subject, claimed to have counted sixty-three different notes in the repertoire of cat miaows, for example, acknowledging that it did take time and practice to do so.

Nevertheless, a large number of these miaows can be detected by the untrained human ear.

There is the request miaow, when a cat wants to be let in or out, for example; the persuasive miaow, a low and coaxing sound made when it wants something it doesn't ordinarily have; a distressed miaow, which is sharp, loud and easily recognizable, but odd in that it is uttered only at the start of a distressing situation. There is also the responsive miaow, in recognition that its name has been called, as well as a greeting miaow.

The purring sound so familiar to cat lovers can also have more than one meaning. Although it is usually interpreted as a sign of a contented cat, it has also been determined that cats purr when experiencing severe pain. And while it may be difficult to tell whether a cat is purring because it is happy or because it is in pain, the mechanisms of its purring are even more mysterious as scientists still do not know exactly how cats make this unusual and distinctive sound.

The extensive repertoire of feline vocal sounds is made more complicated by the fact that the domesticated cat maintains two different sets of vocabularies, one for the wild and one for home. It will instinctively know the appropriate growls, howls and mating calls necessary for life in the wild, even if it has never or never will live outside its domestic environment. On the other hand, it also employs the sounds of a kitten, which in the wild would be for its mother while it was young and then dropped once it became an adult; these kitten sounds are simply elaborated upon for the cat's human owner who fulfils the role of mother through its adult life.

Although an ability to understand the variety of sounds a cat can make is helpful and indeed desirable, it is not necessary to recognize the meaning of them all. One can usually deduce what a cat is trying to say from the combination of signals it gives, including not only its vocal sounds but its use of body language.

An upset cat, after a scolding for example, will often sit at a distance from its owner, with its back to him or her, and will ignore all attempts to be comforted. This is rarely a dignified sulk resulting from a wounded ego, as commonly thought, but rather a way of cutting off the threat the owner has imposed upon the cat with harsh words and an accompanying glare. Because a cat feels intimidated by such hard staring, not to mention the owner's aggressive tone of voice and towering size, its natural reaction is to turn away but without running away like a coward. This response '. . . has a double effect: it reduces the fear in the cat itself and enables it to stay where it is' writes Desmond Morris in his book, *Catlore*. 'It also prevents any counter-staring by the cat, which would spell defiance and possibly provoke further hostility.'

A cat that is feeling friendly, however, will let you know by winding around your legs and purring loudly, often a form of territorial marking. The rubbing of its head against a person's leg or hand actually stimulates the temporal glands on either side of the cat's forehead. Sometimes this causes a very tiny and rarely noticeable secretion from the gland which is left on the spot that was rubbed so the cat can recognize its 'friend' later.

As far as the cat's ability to understand our body language is concerned, because it is a keen observer of all things, including our behaviour, it can usually read our moods and messages quite well. Its excellent visual abilities, hearing and other senses, refined after centuries of hunting, make it an especially receptive if not always demonstrative animal to relate to. Many cats are able to sense when a person is sad or ill, for example, and will stay near by, ready to offer comfort.

In the case of cats born deaf, which occurs most commonly in white cats with blue eyes, they can rely on a person's visible body language to interpret message or mood. They are also able to understand a situation by sensing the vibrations a person or event generate, and they can react to them accordingly.

With their excellent sensory abilities, cats might even be

better at comprehending our messages and moods than we are theirs. The cat writer Paul Corey certainly thought so, proclaiming in his book *Do Cats Think?*, 'I know my cat companions understand my language better than I understand theirs. Although their brains are smaller than mine, I can't help but interpret this achievement of theirs as a show of superior ability of some kind – a superior intelligence . . . or something important to know more about.'

SENSITIVITY

While we may never match the cat in the use of sensory skills to communicate, we can try to be as sensitive to our cat's feelings as they seem to be to ours. Their feelings are expressed in the way they behave and, by taking note of their actions, we can interpret their likely emotional state.

Cats are very possessive, for example, and can become extremely jealous of a favourite chair, pillow or toy if someone tries to share it or take it away. They may also have particular difficulty in accepting new members to the household. New pets or babies should be introduced to the cat gradually, ensuring that every effort is made not to make the cat feel neglected. Signs of jealousy in a cat include the refusal to eat and clean itself properly, as well as sulking.

Cats are also sensitive to being laughed at and this will show in a saddened and uncomfortable expression. And when giving your cat a command, it is important to remember that cats hate to be ordered around and spoken to in a dictatorial tone. It is usually more successful to speak to your cat in a firm but kind tone and to save your stronger tones for more important and serious occasions when strict commands are called for.

Companionship is just as important to cats as it is to humans and a lonely cat may react with a loss of appetite and be lethargic. A cat needs time with its owner, or the companionship of a new pet. Some cats need more than just companionship and crave the undivided attention of their owner or others. They might ask for affection by jumping into their laps or, more discreetly, by rubbing against their legs or simply by sitting near by. If they aren't satisfied with the attention they receive, such cats are likely to develop severe depression.

As naturally curious creatures, cats are happiest when there are enough diversions in their environment to capture their interest. A bored cat will often turn to tearing holes in clothing or in household furnishings, acting depressed, grooming excessively or losing interest in everything around it. In this case, the owner should try to stimulate the cat's interests again with new games or with a new pet.

Lastly, cats are sensitive to loud noises as well as to dramatic experiences which can sometimes traumatize them. The household noises of vacuum cleaners, washing machines and stereos played at a high volume can be frightening to cats, especially to kittens who may not yet be accustomed to them. More critically, shock can develop in cats after an episode of rough handling, moving house, a fight with another animal, mistreatment by children (or adults) or a particularly painful visit to the vet. Initial signs of shock can be detected when the cat's normal habits of hygiene are broken, such as the discontinued use of its litter box. In more severe cases, the cat may become listless and morose, suffering from loss of appetite and a higher than normal pulse rate. Owners should give cats suffering from shock as much reassurance as possible to nurse them back to health, seeking professional care if necessary.

OUR ATTITUDES TOWARDS CATS

Our feelings for our cats and our attitudes about keeping them as pets are extremely important factors in the success of an owner–cat relationship. Some of the qualities that cats are especially appreciated for include their appearance, their independence and unpredictability, their companionship, intelligence, intriguing and mysterious character, and their playfulness and charm. Characteristics which are less admired by cat owners tend to include their cruelty to prey, their ability to stare and make one feel uneasy, and their moodiness.

The nature of the cat is indeed quite complicated, and it is summarized beautifully by Sidney and Helen Denham in a passage from their book, *The Siamese Cat*; 'The ingredients the Creator chose to mix when He decided to make the first Siamese have been given as the grace of a panther, the intelligence of the elephant, the affection of the lovebird, the beauty of the fawn, the softness of down and the swiftness of light.' The quote refers to the Siamese but could easily apply to the nature of all cats.

Yet, as every cat owner surely knows, not all cats are the same and it is their individual and unpredictable character which makes them such special and enjoyable creatures. Likewise, not all owners are the same and their attitudes towards cats can vary greatly.

A good indication of the enthusiasm a cat owner has for his

or her cat is the name that the owner chooses. Some cat therapists believe that cats are sensitive about their names and that choosing a name for one's cat is not a trivial exercise.

In his book *Charles, The Story of a Friendship*, Michael Joseph argued that the naming of a cat is a reliable guide to the appreciation an owner has for it. 'You can be reasonably sure,' he writes, 'that when you meet a cat called Ginger . . . or merely Puss that his or her owner has insufficient respect for his cat. Such plebian and unimaginative names are not given to cats by true cat-lovers.' He goes on to cite the example of a man who christened his cat 'the most noble the Archduke Rumpelstilzchen, Marcus Macbum, Earl Tomlefnagne, Baron Raticide, Waowhler and Scratch', adding that the owner probably shortened this to 'Rumpel' on a day-to-day basis, but the notion of such a name was what mattered.

Renaming a cat is sometimes recommended by cat therapists as a method of treatment for certain disorders. Cat therapist Carole Wilbourn of New York, for example, believes that cats can be affected by their names and claims to have 'transformed a very nervous Siamese' by giving it the name Spencer Tracy.

Of course, not all cats need grand titles to feel good about themselves and confident in the attitudes of their owners. And in actual fact, the naming of a cat is not always an accurate reflection of the appreciation an owner has for it. The eighteenth-century writer, critic and cat lover Dr Samuel Johnson, for instance, simply named his beloved cat Hodge and certainly no evidence exists of a lack of respect or affection on either side of that well-documented relationship.

OUR RESPONSIBILITIES AS OWNERS

Whatever personal attitudes an owner has towards his or her cat, it is most important that the owner understands the responsibilities in keeping a cat as a pet. Animal behaviourists and cat psychologists have determined that the role of the owner in today's domestic relationship with the cat is very much like that of mother for cats raised in the wild. The difference lies in the fact that owners fulfil the role of mother throughout a cat's entire lifetime whereas, in the wild, cats are only mothered for a few months until they can lead independent lives of their own. By perpetuating this stage of a cat's life, our domesticated cats can afford to relax and be sociable as we take care of life's basic necessities.

Cats brought up in the wild enjoy this period briefly, then spend the rest of their lives concentrating on survival. 'We not only provide food when the young adult cat would be expected by its mother to find its own, but we also allow the cat to demand physical attention, warmth and affection from us as only its mother could provide,' writes cat psychologist Peter Neville in his book *Do Cats Need Shrinks?* 'We offer the same security in our laps as the kitten knew when suckling or lying next to its mother.'

An understanding of our role as mother in the relationship will help us to recognize what our cat's needs are and what it expects us to provide for it. At the very least, we must be able to provide for our cat's most basic needs such as diet, grooming and care when it is ill.

A cat's natural diet if it lives in the wild consists primarily of the small prey it catches by hunting mice and other tiny rodents. Cats remain quite healthy on such a diet, and scientists and researchers in the pet food industry have conducted exhaustive tests on pre-packaged foods for the domestic cat to determine the best approximation of minerals and nutrients. Looking at the average mouse, it contains about 70 per cent water, 15 per cent protein, 10 per cent fat and 1 per cent carbohydrate.

Ensuring that your cat has plenty of drinking water is important, especially if it is fed on dried, packaged cat food which does not contain as much water as natural food. Protein can be found in many meats which are popular with cats, as can the fat they need in their diet. Carbohydrates can be digested only if they have been cooked – for example, boiled rice or bread.

Grooming techniques will differ according to the length of a cat's hair. Short-haired cats need grooming just once a week on average, with an easy routine of brushing the coat to stimulate circulation and gather loose hairs, followed by another brushing with a fine comb to remove any tinier bits of dead skin or hair left behind. Long-haired cats require more frequent grooming, usually a daily brushing with the occasional use of a grooming powder to absorb grease from the cat's hair. Both types of cat should have their eyes, ears and other areas inspected on a weekly basis.

To recognize when a cat is ill, one must know how it looks and behaves when healthy. It will have alert and clear eyes with shining pupils that react quickly to changes in light. The eyes should not be cloudy nor water excessively. The ears should be clear of discharge and perfectly clean. The body should be firm and lean, but with no ribs or other bones visible. A healthy cat's coat will be smooth and shiny, with the hairs lying down _lose to the skin in short-haired varieties and slightly fluffed up in long-haired cats. The cat should be alert and curious when it is awake, enjoying physical exercise, regularly scheduled meals and engaging in frequent sessions of self-grooming.

A cat owner is also responsible for giving his or her cat the amount of attention it might need. The more interest an owner expresses in the cat, the greater chance there is that the cat will develop a warm and loving personality. Indeed, this is an even more important factor in determining the character of the cat than any personality traits it might have inherited from its parents.

Lastly, an owner must remember that although cats have been domesticated for centuries, they still maintain a strong desire to hunt, a natural part of their instinct to survive. As hunters, cats will often bring in dead or only half-alive rats or birds meant as presents for the owner. Cats will fully expect to be praised for their effort in these cases and can take great offence at being reprimanded. Praise them in acknowledgement of the 'prize' they have given you and berate them, but not too severely, only if the captured animal is still suffering.

A cat which is kept indoors all or most of the time will need other means of expressing its instinct to hunt. Owners can devise games which challenge and tease their cat. Most cats perk to attention when they hear the soft rustling sounds of paper or fabric, for instance, and can be mesmerized by the noise. Such sounds are very similar to the rustling noises their ancestors listened for in the dark as semi-nocturnal hunters and offer the cat a chance to exhibit its prowess – even if it just attacks a blanket!

COMPATIBILITY

The success of the relationship you have with your cat will also depend on your personalities and on the type of lifestyle you each prefer. Hopefully, you and your cat will be compatible in these respects. While most cats are extremely adaptable to almost any kind of household and owner type, some exhibit strong character traits which may be difficult for owners to handle.

Cats with high energy levels, for example, may be hyperactive and disruptive to one's routine, especially if kept indoors. Such a cat might become burdensome if it turns every activity into a game and will not sit still long enough for its owner to enjoy a little peace and quiet. In this case acquiring another cat as a companion might be the solution. Conversely, some cats prefer to spend most of their time alone and find it annoying if their owner pays them too much attention. This may be frustrating for the owner who would like the companionship of a cat, in which case, again, the addition of a more affectionate cat to the household might be the answer.

Cats that are especially vocal can drive certain types of owners to despair with their incessant miaowing. There is little that can be done to improve the relationship between this type of cat and its owner and, in such cases, it is often better to give the cat to someone else who does not mind a talkative cat. He or she would be more likely to give the cat the attention and care it needs than an owner who has become exasperated with it.

Other cats are particularly curious or mischievous and can irritate certain owners by hiding objects, tearing into rubbish bags and breaking up household items in the name of entertainment. In this respect, the degree to which a cat follows the rules of the house can affect the nature of the relationship with its owner. Here the intelligence of the cat comes into play, as does the method in which household rules are conveyed to the cat by its owner. Rewarding good behaviour is just as important as punishing undesirable behaviour in reinforcing a rule. One should also reprimand a cat for a misdeed as soon as possible after the act was committed, so that the cat can make the connection between the act and the punishment.

A cat's social behaviour is also important to owners, especially if the owner has children or a large number of visitors to the household. Cats can be trained from a young age to accept the presence of children or strangers and the correct ways of behaving. If a cat has not learned to be well-mannered though, an owner can either try to teach it to behave or simply ensure that the cat is kept separate.

Furthermore, just as married couples are usually happiest when both partners share the same type of body clock, so pets are best suited to owners who maintain a compatible daily schedule. A cat that wakes up early and makes a noise will not enthral an owner who would prefer to sleep a few more hours. Nor will a morning cat be happy with an owner who stays up late at night, making it difficult for the cat to fall asleep. In such cases, cats will usually adapt by finding some spot in the house where they can attempt to sleep, and they are also blessed with an ability to block out irrelevant sounds while sleeping, although they may not sleep as well. Owners with cats who are noisy at inappropriate hours might try keeping the cat confined to another part of the house during that time.

REWARDS OF A HAPPY RELATIONSHIP

If the personalities and lifestyles of both the cat and its owner are suited to one another, and each understands its responsibilities and expected behaviour, a successful relationship between the two can develop which brings with it a number of wonderful rewards. Cats offer their owners many emotional benefits which in turn help our physical well-being.

'It often happens that a man is more humanely related to a cat or dogs than to any human being,' wrote the poet Henry David Thoreau in the nineteenth century. Indeed, this is still the case today. Cats serve as an emotional outlet for many people throughout the world. A recent University of Pennsylvania study concluded that 37 per cent of American cat owners talk to their cats regularly about the events of the day or confide in them about more personal matters. Although they may not be able to offer any advice in return, the intelligent and alert nature of most cats makes them reassuring to talk to and their silence contributes to a certain aura of wisdom which may be imaginary, but is none the less comforting.

Cats can also make us feel good about ourselves because we are able to provide for them, helping us feel needed and important. We tend to develop a healthy attachment to our cat as we take on the role of its mother, protecting and caring for it as best we know how.

Cats are often comical and can make us laugh at what they get up to or simply by their expressions. They bring us presents, laying their catch or other gifts at our feet. They are beautiful creatures and pleasing to watch as they move gracefully. Because of their uncanny ability to register time, many cats will know when their owner is due home and will wait to welcome him or her. They are especially treasured by owners who live on their own, making them feel loved.

The emotional benefits of owning a cat have positive repercussions for one's physical and psychological well-being. Most people find holding a cat relaxing, with the warmth of its body, the soft touch of its coat and the calming sound of its purr contributing to the effect. A cat's care-free attitude around the house can often also rub off on an owner.

Stroking a cat can be therapeutic for many humans, alleviating stress and lowering one's blood pressure and heart rate. A team of researchers at the University of Maryland, for instance, found that heart-attack victims who owned a pet had a greater probability of surviving their first year than those patients who did not own a pet. Perhaps the acquisition of a pet will be prescribed by doctors one day as an aid to recovery!

Cats are especially well placed as pets in today's modern world. Although their natural environment is the wild, cats are extremely adaptable and can generally learn to live quite happily inside for their entire lives. Most cats are willing to substitute a litter box for the great outdoors, for example, and most will invent games and imaginary chases to replace the hunting that they are instinctively interested in and naturally equipped for.

Cats are also becoming more attractive to keep because of their relatively low maintenance requirement. They are fastidious and like to keep themselves clean, doing their own grooming daily. They do not need to be walked, they can be trained to use their litter box and are usually able to get enough exercise within the confines of their home. Cats are usually quiet, which is helpful to busy owners who may not spend much time at home except to sleep. Economically, cats are easier to keep as they eat less than dogs and are not as expensive to care for.

Some people, however, dislike the fact that the cat is so independent and self-sufficient, and many dog-lovers believe that the cat takes its owner for granted, greedily receiving free food and shelter and giving little back in return. But dog-lovers tend to prefer a pet that will show them affection on command. This is perfectly reasonable as long as the dog owner has the extra time and resources to give his or her dog the proper care it requires.

Increasingly in today's society, though, this is not the case and the practical advantages of keeping a cat as a pet are becoming more important. Cat lovers might add that cats make better pets *because* they are not completely subservient to their owners, and that their independent nature is actually a strength. To a cat lover, this quality gives the cat a more interesting, entertaining and intriguing personality and makes it more challenging and stimulating to keep as a pet. And although intelligence is a less important factor in the success of a relationship between a cat and its owner, it is likely that any cat lover will claim that cats are also more intelligent than dogs.

These personality traits of the cat, coupled with the emotional and physical benefits of its companionship, make the cat a wonderful housepet in today's modern age. The cat is remarkably well suited to a life shared with man, a life made richer by its presence.

VII
REFERENCES

Aberconway, Lady C., *A Dictionary of Cat Lovers*, Michael Joseph, 1968 (first edition 1949).

Anderson, Janice, *Cat Calls*, Guinness Publishing, Enfield, Middx, 1991.

Angel, Marie, *Catscript*, Pelham Books, London, 1984.

Angus, Vivienne, *Know Yourself through Your Cat*, Souvenir Press, 1991.

Armour, Robert A., *Gods and Myths of Ancient Egypt*, The American University in Cairo Press, Cairo, 1986.

Baker, Stephen, *How to Live with a Neurotic Cat*, Grafton Books, 1987.

Beadle, Muriel, *The Cat*, Simon & Schuster, New York, 1977.

Butler, E. and Madsen, P., *Test Your I.Q.*, Pan Books, 1983.

Chaucer, Geoffrey, *Canterbury Tales*, translated by Nevill Coghill, Penguin Books, 1951.

Corey, Paul, *Do Cats Think?*, Castle Book Sales Inc., Secaucus, New Jersey, 1977.

Denham, Sidney, *Cats between Covers*, H. Denham (London), 1952.

Foster, Dorothy, *In Praise of Cats*, Musson Books Co., Toronto, 1974.

Joseph, Michael, *Charles, the Story of a Friendship*, Michael Joseph, 1943.

Kirk, Mildred, *The Everlasting Cat*, Faber and Faber, 1977.

Leman, Jill and Martin, *The Perfect Cat Anthology*, Pelham Books, 1983.

Lillington, Kenneth, *Nine Lives*, André Deutsch, 1977.

Morris, Desmond, *Catlore*, Jonathan Cape, 1987.

Neville, Peter, *Do Cats Need Shrinks?*, Sidgwick & Jackson, 1990.

New Encyclopaedia Britannica, vol. 1, Encyclopaedia Britannica Inc., Chicago, 1988.

Page-a-Day Calendar, 1991, Workman Publishing Co., New York, 1990.

Pedigree Petfoods, *Know Your Cat*, U.K., 1989.

Pielou, Adriaane, '100 Ways to Cure a Cat', *Mail on Sunday*, 11 November 1990.

Pond, Grace, *The Complete Cat Encyclopedia*, Heinemann, 1972.

Pond, G. and Sayer, A., *The Intelligent Cat*, Davis-Poynter, 1977.

Pugnetti, G., *The MacDonald Encyclopedia of Cats*, MacDonald, 1983.

Reid, Beryl, *The Cat's Whiskers*, Ebury Press, 1986.

Sayer, Angela, *Encyclopedia of the Cat*, Octopus Books, 1979.

Schneck, M. and Caravan, J., *Cat Facts*, Stanley Paul Publishers, 1990.

Sitwell, Osbert, *The True Story of Dick Whittington*, Home & Van Thal (London), 1945.

St George, E. A., *Ancient and Modern Cat Worship*, Spook Enterprises (London), 1981.

VIII
WHAT WERE YOUR RESULTS?

The author would like to produce a sequel to this book which evaluates the I.Q.s of owners and their cats from different countries around the world. Results will be assessed to determine which countries have the brightest cats, for example, and the most fanatical owners, which cat breeds are the cleverest and what impact, if any, age and sex have on the scores.

A collection of clever and blissfully ignorant cat stories will also be included. If you have a story or anecdote about your cat, which demonstrates its high level of intelligence – or lack thereof – please send it in as it could be included in the book. If so, the names of you and your cat will be kept in the story, unless you prefer to remain anonymous.

THANK YOU FOR YOUR CONTRIBUTION

Results

PLEASE PRINT ALL YOUR INFORMATION

The Definitive Cat I.Q. Test

NAME OF CAT:

NAME OF CAT:

AGE: SEX:

AGE: SEX:

BREED/DESCRIPTION:

BREED/DESCRIPTION:

TEST SCORE: CAT I.Q.:

TEST SCORE: CAT I.Q.:

The Definitive Cat Owner I.Q. Test

NAME OF OWNER:

NAME OF OWNER:

AGE: SEX:

AGE: SEX:

NO. OF YEARS YOU HAVE KEPT A
CAT:

NO. OF YEARS YOU HAVE KEPT A
CAT:

TEST SCORE: OWNER I.Q.:

TEST SCORE: OWNER I.Q.:

CITY/COUNTRY:

CITY/COUNTRY:

IF YOU ARE ENCLOSING AN ANECDOTE OR STORY,
PLEASE INCLUDE YOUR ADDRESS:

Intelligent or Blissfully Ignorant Cat Story

☐ My cat and I would like to be named in this story, if included.

☐ My cat and I would prefer to remain anonymous in this story, if included.

(PLEASE TICK THE APPROPRIATE BOX.)

_____

_____

_____

_____

_____

_____

_____

_____

_____

_____

_____

_____

_____

_____

_____

_____

Please return to Melissa Miller,
Definitive I.Q. Test for Cats and I.Q. Test for Cat Owners,
c/o Penguin Books, 27 Wrights Lane, London W8 5TZ.

IX
SAMPLE POOL OF DOMESTIC CATS
LIST OF PARTICIPANTS

NAME OF CAT	BREED/DESCRIPTION	SEX	NAME OF OWNER	CITY
Adolph	Black and White Moggie	F	Maureen Day	London
Agnes	Tabby	F	Jilly Cooper	Stroud, Gloucestershire
Alex	British Blue	M	Peter and Pam Smith	London
Arnold	Long-Hair Tabby	F	Karen Thrumble	Edinburgh
Babs	White with Black Spots	F	J. D. and Christy Kennedy	Dallas, Texas, USA
Barney	Grey and White Moggie	M	Bob and Kim Stares	Sawbridgeworth, Hertfordshire
Bimbo le Mog	Black Moggie	F	Celia Haddon	London
Bonny	Tortoiseshell Tabby	F	Tricia Scanlon	Orpington, Kent
Boris	Black and Mean Moggie	M	Laura and Mark Kerns	Chelsfield, Kent
Buster	Ginger Tom	M	Sarinder and Ann Singh	Edinburgh
Cat-Purry Fatty Monster	Brown Burmese	M	Sybil Ann Chick	London
Champagne	Cream Colourpoint	M	Laura Miller	Atlanta, Georgia, USA
Charlie Mace	Black and White Heinz 57	M	C. A. Mace	London
Cinnamon	Havana	F	Charlotte Carson	Tangley, Hampshire
Cleo	Black and White Domestic Short-Hair	F	Gillian Fryer	London
Doris	Black and Brown Tabby	F	Paul Clark	Edinburgh
Elvis	White Las Vegas Fluff Ball	M	Laura Boomer	London
Figaro	Chocolate Burmese	M	Carol Smith	London
Fred	White Tom	M	Jenni Edwards and Bryan Laidlaw	Edinburgh
George	Blue Colourpoint	M	Bob and Kim Stares	Sawbridgeworth, Hertfordshire
Hazel	Oriental Tabby	M	Charlotte Carson	Tangley, Hampshire
Henry	Cute and Cuddly Moggie	F	Laura and Mark Kerns	Chelsfield, Kent
Hobs	Moggie	F	Mr and Mrs Peter Warrington	Beaconsfield, Buckinghamshire
Horace	Tabby Tom	M	Debbie Stuart	London
Inky	Black Moggie	F	Maureen Day	London

NAME OF CAT	BREED/DESCRIPTION	SEX	NAME OF OWNER	CITY
Janet	Black Moggie	F	Octavia Wiseman	London
Jelly	Black and White Moggie	M	Eileen, Frank and Tom Orford	London
Jerry	Tabby	F	Frances Taylor	Farnham, Yorkshire
Kai	Blue Smoke Persian	M	Laurie and Daryl Sartain	Dallas, Texas, USA
Kip	Blue Smoke Persian	M	Laurie and Daryl Sartain	Dallas, Texas, USA
Kippy	Tabby	F	Jean and Ian Haddow	Edinburgh
Kiri	Black Heinz 57	F	Elaine Briggs	Edinburgh
Kizzy	Burmese/Moggie	F	Jacquie and Mabel Edwards	Edinburgh
Lion	Moggie	F	Jane Butcher	San Francisco, California, USA
Liza	Tortoiseshell	F	Pamela Mackrell	London
Louis	Black and White Moggie	M	Shona McLeod	London
Lucy	Black and White Moggie	F	Peta Dempsey	Edinburgh
Lulu	Black Moggie	F	Nick White	London
Macavity	Angora	M	Amy and Georgina Agnew	London
Michaela	Russian Blue	F	Fred Bennett	Chorleywood, Hertfordshire
Mingaladon	Tonkinese Seal Point	F	Sylvia Nuttman	Beckenham, Kent
Minty	Moggie	F	Mr and Mrs Peter Warrington	Beaconsfield, Buckinghamshire
Missy	Short-Hair Tabby	F	Jane Keys	Edinburgh
Muff	Moggie	F	Phil and Amanda Britton	Edinburgh
Muffin	Calico	F	Nancy Solomon	Dallas, Texas, USA
Mullian	Tortoiseshell	F	Mrs Anna Connor	Wheatley, Oxfordshire
Mungo	Black and White Domestic Short-Hair	M	Jack Crossley	London
Mungojerry	Tabby	M	Amy and Georgina Agnew	London
Nimrod	Black British Short-Hair	M	Sylvia Nuttman	Beckenham, Kent
Otis	Black Alley Cat	M	Laura Boomer	London
Pepe	Blue Persian/Heinz 57	M	Elaine Briggs	Edinburgh
Puff	White Tabby	M	Pamela Mackrell	London
Rocky	Javanese	M	Frank Miller	Dallas, Texas, USA
Rudolph	Russian Blue	M	Fred Bennett	Chorleywood, Hertfordshire
Samson	Ginger Domestic Short-Hair	M	Jack Crossley	London
Sidney Vicious	Black and White Long-Hair	M	Christy and Jerry Kennedy	Dallas, Texas, USA
Smokey	British Blue	M	Jane Keys	Edinburgh
Spats	Moggie	M	Phil and Amanda Britton	Edinburgh
Spindle	Brown Burmese	F	Jane and Jacqui Morris	Hainault, Essex

NAME OF CAT	BREED/DESCRIPTION	SEX	NAME OF OWNER	CITY
T. C. Mace	Black and White Heinz 57	M	C. A. Mace	London
Tiger	Short-Hair Tabby	M	The Verlander Family	London
Tigger	Classic Tabby	M	Sybil Ann Chick	London
Tiggy	Black and Grey Domestic Short-Hair	M	Alwyne Kerns	Shrewsbury, Shropshire
Tiggy	Black Moggie	M	Mrs A. Winsor	Kenley, Surrey
Toby	Black and White Moggie	M	Sybil Ann Chick	London
Tom	Ginger	M	Frances Taylor	Farnham, Yorkshire
Toots	Tabby	F	Mr and Mrs David Kimmond	Kennoway, Fife
Topsy	White Moggie	F	Sybil Ann Chick	London

X
READERS' RESULTS

MANY THANKS to all the readers who sent in their I.Q. test results and entertaining cat stories. I enjoyed reading each one! This chapter gives an analysis of those results and includes some of the best stories I received about cat intelligence and cat ownership. The results represent over a hundred cats and their owners, over half of whom live in the U.K. (Stories and results received after 1 May 1994 will be included in future updates to this chapter.)

The range of I.Q. scores that I received has been quite wide – both for cats and cat owners. On a possible scale of 60 to 180 the cat I.Q. scores range from 70 to 156, and cat owner I.Q.s range from a Practical owner's 79 to a very Fanatical owner's 152!

But what's most interesting to see is whether I.Q. scores vary by type of cat or by type of owner. For example, do moggies have higher I.Q.s than pure breeds? Are long-haired cats brighter than short-haired cats? Is an owner's I.Q. likely to increase the longer he or she keeps a cat? The graphs and tables in Part I of this chapter contain the answers to these and other questions. In some cases, there are clearly discernible trends.

Part II contains my selection of stories and anecdotes about cats who are Blissfully Ignorant or Extremely Intelligent, as well as stories that offer a particular insight into the cat–owner relationship. Although some are reproduced only in part, all the stories have been kept in the contributors' own words.

**PLEASE KEEP YOUR STORIES AND RESULTS
COMING IN FOR FUTURE UPDATES!**

Part 1 Cat and Owner I.Q. Comparisons

1. Cat I.Q. Analysis:
Short-haired Cats versus Long-haired Cats

Are Short-haired Cats Cleverer Than Long-haired Cats?

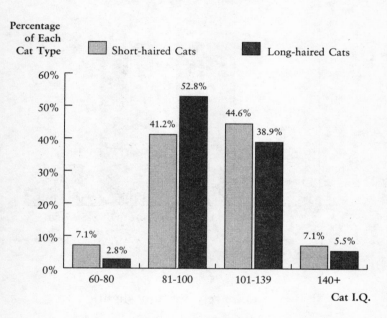

On average it seems that they are . . .

AVERAGE CAT I.Q. SCORE

Short-haired Cats
106.1

Long-haired Cats
101.5

2. *Cat I.Q. Analysis: Pure Breeds versus Moggies*

Are Pure Breeds Brighter Than Moggies?

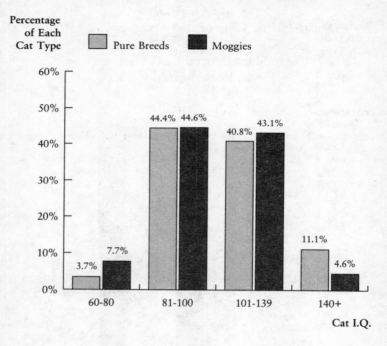

In this sample, yes, but only slightly . . .

AVERAGE CAT I.Q. SCORE

Pure Breeds
105.6

Moggies
103.8

3. *Cat I.Q. Analysis: Male Cats versus Female Cats*

Who are More Intelligent – Male Cats or Female Cats?

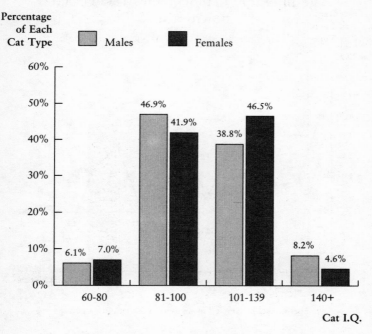

Percentage of Each Cat Type

Males Females

It depends how you look at it!

More female cats have an I.Q. above 100 . . .

(51% versus 47%)

But males have a higher average score:

AVERAGE CAT I.Q. SCORE

Male Cats
104.6

Female Cats
101.8

4. Cat I.Q. Analysis: Age and Cat I.Q.

There appears to be little correlation between age and cat I.Q., though the overall trend is downward . . .

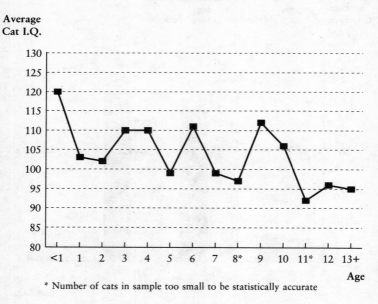

Average Cat I.Q.

Age

* Number of cats in sample too small to be statistically accurate

5. *Cat Owner I.Q. Analysis*

An owner's I.Q. is likely to increase the longer he
or she keeps a cat . . .

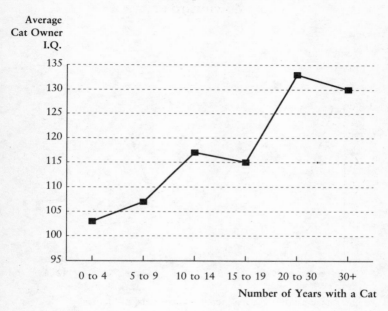

But, in most cases, cats still have the upper paw . . .

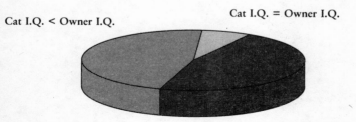

Part II Entertaining Cat Stories

SECURELY UNDER THE PAW...

I have long been suspicious that our feline is the most intelligent being in the household. She invariably comes off best in the daily battle of wits which is fought between us. If I am so smart, how is it that she spends her day asleep in the most comfortable chair, while I have to work to feed her?

... I now realize that the only way to get any peace is to settle the question of our cat's I.Q. once and for all. Certainly some of the traits mentioned seemed very familiar. Our manipulative moggy exhibits considerable audio skills. She can arrange to be unavailable when the words V–E–T and C–A–T–T–E–R–Y (now known in our family as the V-word and the C-word) are mentioned, and can distinguish the particular sound of a rustling crisp packet at thirty paces in a Force 8 gale.

She shows interest in animals on the television. Faced with a new kitten, she would probably lure it to an untimely end. She packs herself in suitcases if one of us so much as suggests going away. Her subtle approaches to gaining attention when we are asleep border on sheer genius. (How does a cat *know* that one's bladder is full before jumping on it?) ...

Yours, securely under the paw, J. Laker, Middlesex

A MAN WHO SHALL REMAIN ANONYMOUS WRITES...

I hope the cat does not prove more intelligent than me (or my wife – much more likely!)

IN 'LIEU OF A LOO', NECESSITY IS THE
MOTHER OF INVENTION...

Siam, a Siamese, was accidentally locked in my bedroom one evening when I went out, and I didn't return home until very late that night. When I came back and realized he'd been in the bedroom all that time, my first thought was, Oh, no, I hope he hasn't ruined the carpets!

But I needn't have worried. Siam had found some newspapers I'd left on the floor; he shredded them up, made a pile of shreds in the corner of the room, and used *that* as his 'litterbox' for the night!

– Martha Foster Skyles, Dallas, Texas

EASY RIDER

We have to watch Toffee as she loves travelling in cars, so all visitors' cars have to be checked in case they have a stowaway. I once reversed the car out of the garage and was halfway down the road when I was flagged down by a neighbour who told me I had a cat on my roof. I discovered Toffee clinging to the sunroof, purring loudly and enjoying the ride. Needless to say, we now check the outside as well as the inside of the car!

– Sue Blyth, 'Miaow', Mensa Cat Group Newsletter

CLEVER PERDITA CHARMS
HER WAY IN...

Our cat adopted us last February (1992). We named her Perdita, which is Latin for lost. Her escapades include chasing a frog into the house and under the piano, and turning on the bedroom light at 3 a.m. by attacking the pull cord.

We fully expected her to pull all the tinsel, etc. off our Christmas tree, which she does. But we were surprised when she climbed right up to the top and sat there like a furry fairy.

But her cleverest achievement is undoubtedly getting invited into the home of four people, two of whom are allergic to cats and the other two who didn't really like cats, simply by sitting on the doorstep looking pathetic and pretty.

— Irene Boyd, Stockport

IT'S EASIER TO KNOCK ...

Our son's cat, Mittens, is a sweetly pretty tabby with a punk-like ginger tuft on the top of her head ... About a year ago Mittens discovered that if she stands on her back legs and stretches as far as possible, she can just reach the letterbox and rattle it with her front paw. So now, every time she wants to come in, she knocks on the door!

After all, what's the point of clambering over a six-foot gate and climbing through the cat flap in the back door when you can get a human to open the front door for you? And *we* are supposed to be the highly intelligent ones?

— Carrie Howse, 'Miaow', Mensa Cat Group Newsletter

I HAVE 'WAYS' OF MAKING YOU MOVE...

At the time we had two cats – assertive, aggressive Tamsin and gentle Kiwi. John sat by the fire. Tamsin shouldered Kiwi aside and jumped on to John's knees, settled, purring, tail hanging down. Kiwi 'thought' for a moment, then gently crept forward and bit Tamsin's tail. Tamsin started up, digging his claws into John who responded by pushing Tamsin to the floor.

Immediately Kiwi leapt on to John's knee and settled. It could have been chance, but to me – the observer – the speed of the event left me in no doubt that Kiwi had actually worked it out . . .

– Trudie Hodge, Plymouth

INTRODUCING THE INCREDIBLY NIMBLE PUFF...

. . . Puff remains, as ever, a miracle of grace and loveliness . . . as well as strangly boneless; she has the disconcerting ability to become practically two-dimensional, as she demonstrated recently when she decided to sit on top of the books in one of the bookcases . . . in the six-inch gap between the books and the upper shelf. The expression of idiotic complacency on her whiskers cannot be adequately described . . .

– Hilary Broadribb, 'Miaow', Mensa Cat Group Newsletter

WARDROBE–HO!

I used to have an old wardrobe whose door wouldn't shut properly. I kept woolly jumpers on the shelf at the top. Wee Cat, of course, explored this immediately (she always does with new things; Dennis waits till she's checked them for bombs, snakes, etc., then evicts her). Soon she was asleep in the jumpers.

One night I saw how she got there. Jumped on to the dressing-table, then on to the top of the wardrobe, then gently pushed the door open just a wee bit with her front paws. Then carefully stood with all four paws on the top of the door, turned round, and hopped into the top shelf and snuggled up.

Dennis was also studying this procedure. So, he leapt on to the dressing-table (skidding, of course, on the mat), leapt on to the top of the wardrobe (dislodging a rolled-up sleeping bag *and* a cat basket), put his front paws on top of the door, and pushed. And kept on pushing till he was stretched to his full and very considerable length, with front paws on the door and back ones on the wardrobe, and an 'Oh, no, I've got it wrong again' look on his face.

Something had to give. It was his back paws. He splatted against the inside of the door, hung precariously by the tips of his front paws for a few seconds, then slid, claws extended, down the mirror inside the door. Wee Cat watched all this in total disbelief . . .

And One More Attempt . . .

A couple of nights later Dennis brought his superb intellect into play and decided to approach the shelf by climbing up my clothes . . . By an amazing feat of twisting and leaping, he managed to get his front paws on the (upper) shelf . . . Unfortunately, he had gripped a sweater, so when he put his full weight on his front paws, instead of getting purchase on the shelf, he pulled all the sweaters, and the very angry Wee Cat, off the shelf and they all crashed to the floor. Since that day (some ten years ago), he has pointedly ignored all wardrobes.

– *Judith Nicholson, Fife, Scotland*

THINGS TO SAY TO A WELL-EDUCATED CAT

Feles mala! Cur cista non uteris?
Stramentum novum in ea posui.

Meaning: Bad kitty! Why don't you use the cat box?
I put new litter in it!

– Latin for All Occasions, Henry Beard
(*Submitted by Maureen Day, 'Miaow', Mensa*
Cat Group Newsletter)

SO THAT'S HOW YOU DO IT!

After I had Simba for a year . . . he wanted to go in and out so much that I fitted a cat flap. One day, he chased a magpie and caught the long tail feathers. He picked one up and proudly ran to the house to show it to us, but as it was crossways, he could not get in through the flap.

He would not come in . . . until we opened the door. I showed him how to drag the feather through the flap and he has never made the same mistake again!

– *John Butler, Wimbledon*

ONCE BITTEN, FOREVER SMITTEN...

I am owned by an intelligent red Persian male called Leo. Well, about six months ago, he had to have a tooth out, so to compensate for his discomfort and lack of dignity, I fed him favourite treats. It wasn't a good idea though, because ordinary tinned cat food was no longer acceptable to my feline friend.

Treats were costing me a fortune. I couldn't believe my good luck when my elderly mother broke the news that Leo had actually devoured a bowlful of ordinary cat food — after he had made sure I was out of sight, of course. I went to bed relieved that my worries were over but at 2 a.m. the sound of a crash woke me. Leo had landed on my radio and, not content with that, he began to abuse the telephone. I finally admitted defeat when he padded all over my head. Bleary-eyed, I scooped some 'treat' into the familiar china bowl. Leo is still being fed a diet of treats.

— Mavis Beckett, Fife, Scotland

A BIT ABSENT-MINDED PERHAPS?

Dennis has ... walked off the end of a mantelpiece (he was looking at the TV at the time), attempted to sleep on a totally unstable pile of books, and somehow (I've never yet understood how) managed to roll himself up in a tablecloth and fall off the table in a kind of cocoon. He's the best cat I've ever had!

— Judith Nicholson, Fife, Scotland

HERE'S AN OWNER WHO KNOWS HER PLACE...

I live with two cats and a dog. I can't pretend that they live with me because they run the house and everything revolves around them ... Where do I fit into the scheme of things? Well, like most cats' lodgers, I'm there to mop up on the odd occasion when there is an accident; I provide the transport and fees for the vet when sickness rears its ugly head; I am a feeding machine, a door opener, a ward and comfortable cushion, and something to lash out at when tempers flare.

Even in my menial position in the household, I adore these animals who can also be so soft and loving, and I know that without them my life would be much the poorer and exceedingly dull. Indeed I am looking forward to becoming a cranky old woman with a house full of cats.

— Sheila Fowler, 'Miaow', Mensa Cat Group Newsletter

MASTERING MODERN TECHNOLOGY

The most amazing thing my cat Tika does is to answer the telephone. I have an extension in both the dining room (wall phone) and a table-top phone in the den. When the phone rings I invariably answer in the dining room. Tika goes into the den and flips the phone out of the cradle and eavesdrops on my conservations. (She doesn't talk.)

— Susan Lalor, Delaware, USA

In order to capture my attention 'after hours', Chester will use his paw to stomp on the pager button on the cordless phone. He rings the pager until I awaken. At first I thought it was a hang-up call until I saw him do it with my own eyes. His usual times are around midnight and 5 a.m.

— Hariann Goldman, Los Angeles

OK, GLOVES OFF

When Bilbo was very young, he used to crawl into my right armpit until I went to sleep. He would then climb up on my chest and proceed to reduce my left hand, usually resting quietly thereon, to raw hamburger. Quickly tiring of these sudden, painful awakenings, I took to wearing an old glove to bed . . . One morning, I woke up to find myself gloveless, though unmauled – I searched everywhere for that glove, but found it only when I pulled the bed away from the wall to tidy it for the day: a soft 'plop' alerted me as the glove fell to the floor. Bilbo had not only managed to ease the glove off my hand without waking me – he had also stuffed it down behind my head, between the bed and the wall.

– *Patricia Halls, Valais, Switzerland*

A SLOW LEARNER

We have owned a tropical aquarium for two years. As soon as it was set up and fish were introduced, Paddy perched on top, paws hanging down, swiping at the fish. Mick didn't seem to notice, until last week – it's only taken him two years. One day he walked past the aquarium and seemed to notice something moving. Since then he is a cat obsessed! He is the one perched on top, paws hanging down, swiping. Poor Paddy has had to find other amusements. He doesn't seem too bothered, as after all he discovered that source of amusement years ago!

– *Linda Williams, Liverpool*

FEARLESS EM TO THE RESCUE

Em (Emily) was fifteen years old. We were living in the forest at our Fire Lookout Tower (my husband's employment) near Mile Zero of the Alaska Highway. One very, very windy day, I was paying a visit to the outhouse. (I left) the door half open ... Suddenly a bear appeared – a black bear – in profile in the doorway. I didn't know what to do. I thought I could frighten him away if I yelled, and banged the half-open door into his face. So I did those things. Did the bear run away? No. He simply turned his face *toward* me, drooling and growling, as if to say 'Ah, *there* you are.' I began to scream – wondering why Ross, my husband, wasn't rushing to my aid. (But) he was about 100 yards away and couldn't hear me because of the wind.

I kept on screaming, the bear just kept staring at me – and then I saw, flying through the air, straight into that bear's face, my Em. My gentle old darling, protecting me. The idea of my cat being injured alarmed me so much that I abandoned caution and fear and hurled myself out the door of the outhouse, nearly colliding with Ross who had finally heard my screams, (had) picked up a long metal pipe, and was brandishing it at the bear like a very long sword. The bear turned tail and ran, Em chasing him to the end of the path. From then on she never let me go to the outhouse without her.

– *Billie Gates, Alberta, Canada*

GROOMING 101

While a kitten, Charlie would invariably lick one paw and then proceed to wash his face with the other (dry) paw! (Fortunately, he outgrew this stupidity.)

– *Pat and Cliff Crader, Cedaredge, Colorado*

AHEM, IT'S MY TURN, NOW

I have a lighted make-up mirror that turns so that you can use the magnified mirror on the back. I spend a lot of time in front of the mirror, putting make-up on, taking it off, etc. When Selena decides that I have spent enough time there and that it's time for me to play with her or feed her, she walks between me and the mirror, pushes it with her paw so that it turns, and sits there until I stop what I'm doing and find out what she wants or needs.

– *Christina Shevalier, Nashville, Tennessee*

DON'T GET UP, I'LL HELP MYSELF

One day I noticed a large number of straws (on the floor) and it looked like my cat Oliver was in heaven with them. I knew I hadn't taken them out (and wondered) how he could have got up into the high drawer (where they were kept).

The next week, I saw him look around and climb up the rungs of the drawers beneath the straw drawer; (he) stuck his nails under the top of the drawer and pulled backwards at a 45-degree angle. Then he climbed in the drawer and brought the straws out. His best trick yet.

— Melissa Calavan, Oakland, California

CAT

Alluringly distant,
Seductively cool,
Disarmingly playful
But nobody's fool.

Endearingly wilful,
Aloof from the crowd,
Bewitchingly haughty,
And famously proud.

This lovable tyrant,
Beguiles and unnerves,
And makes of his master
A minion who serves.

— Tim Hopkins, Luton, Bedfordshire

MOTHER'S HELPERS

My cats seem to have realized when I was pregnant and how to treat a baby. When I brought our daughter home from the hospital, they all gathered round to greet her. Thereafter, everytime she cries they come running to make sure she's OK. If she's upstairs crying, Bunny will come downstairs where I am and meow incessantly until I check on the baby. They are very gentle and protective with her and take their 'sibling' role seriously! (Of course, they're generously rewarded for every effort.)

– *Darlene and Bill Kramer, Kenosha, Wisconsin*

ALARIC, THE MARTYR

Alaric is extremely vindictive if forced to do something he doesn't want to do, such as go to the vet. However, his method of paying me back is to tear one of *his* toys to bits, and arrange the pieces on my pillow. He never destroys anything of mine, Dumb?

– *Sandra Siddall, Sylvania, Ohio*

NOTES FROM THE APPROPRIATELY NAMED, ROLEX . . .

Whenever someone comes over to our house with money, he grabs it out of their hand.

. . . AND THE SPOILED ROTTEN, CAMEO

Cameo loves to be spoiled. She eats in the powder room and will finish half her food, then meow, but only when Vivian is available. Vivian then has to sit on the closed toilet and feed Cameo while telling her how smart and beautiful she is.

– *Ed and Vivian Lichtman, Bensalem, Pennyslyvania*